BASICS OF
Supply Chain
Management

Titles in the Series

BASICS OF
Supply Chain
Management

by

Lawrence D. Fredendall
Ed Hill

The St. Lucie Press/APICS Series on Resource Management

St. Lucie Press
Boca Raton • London
New York • Washington, D.C.

**THE EDUCATIONAL SOCIETY
FOR RESOURCE MANAGEMENT**
Alexandria, Virginia

Library of Congress Cataloging-in-Publication Data

Fredendall, Lawrence D.
   Basics of supply chain management / by Lawrence D. Fredendall and Ed
Hill.
      p. cm. — (St. Lucie Press/APICS series on resource management)
   Includes bibliographical references and index.
   ISBN 1-57444-120-5
   1. Business logistics. 2. Materials managment. I. Hill, Ed, 1947–   .
II. Title. III. Series.
   HD38.5.F738 2000
   658.7—dc21   00-011246
                                                                CIP

© 2001 by CRC Press LLC
St. Lucie Press is an imprint of CRC Press LLC

No claim to original U.S. Government works
International Standard Book Number 1-57444-120-5
Library of Congress Card Number 00-011246
Printed in the United States of America 1 2 3 4 5 6 7 8 9 0
Printed on acid-free paper

Dedication

I wish to dedicate this book to my wife, Elaine Malinowski Fredendall
Lawrence Fredendall

To my wife, Patricia, and our daughter, Kelly, I dedicate this book
Ed Hill

Dedication

Contents

Continuous Improvement
Focused Factory—JIT tool
Simplification
Standardization

Playing "The Beer Game"
There Is a Solution
The Same Solution Will Apply to Any Supply Chain

Supplier Relationships
Supplier Involvement

Quality Function Deployment (QFD)
ISO 9000 Quality Management System
QS 9000
Malcolm Baldrige National Quality Award
Standard Problem-Solving Methods

Cross-Functional Work Team
Self-Directed Work Team
Empowerment
Benefits of Teams

Little's Law
Inventory Management
Types of Inventory
Inventory Policy
Order Quantity Rules
Inventory Classification
Cycle Counting
Materials Requirements Planning

Purchasing Cycle
Supply Problems
Logistics
Integrated Logistics

Internal Data Integration
Advanced Planning Systems
ERP and the Internet

Electronic Commerce on the Internet
Barriers to Using the Internet in Supply Chain Integration
Dell Computer's Factory
MIS Implementation

Preface

S upply Chain Management (SCM) was once a dream, a concept more than a reality, since there were many necessary components of supply chain management that could not be fully achieved. A key barrier to full supply chain management was the cost of communicating with and co-ordinating among the many independent suppliers in each supply chain. An entire supply chain stretches from the creation of raw materials to the deliv-ery of the finished consumer goods. Because firms are involved in many, many supply chains, active supply chain management is practical only for items es-sential to the firm's market success.

Managers are increasingly interested in actively managing their supply chains because of three environmental changes. First, technology has been developed to simplify communication between members of the supply chain. Second, new management paradigms have developed that are being widely shared among supply chain members so that it is simpler for these managers to coordinate their efforts. Third, the development of a highly trained work-force allows employees at each stage of the supply chain to assume responsi-bility and the authority necessary to quickly make decisions and take actions required to coordinate the supply chain.

While the three changes above make supply chain management possible, it is competition in the marketplace that is pushing firms to make SCM a re-ality. Those who master SCM gain a competitive advantage. So, SCM means money. And, SCM means jobs.

For the past 30 years the business world has been inundated by concepts and jargon. These include: Materials Logistics Management (MLM), Just-in-Time (JIT), Materials Requirements Planning (MRP), Theory of Constraints (TOC), Total Quality Management (TQM), Agile Manufacturing, Time Based

Competition (TBC), Quick Response Manufacturing (QRM), Customer Relationship Management (CRM), and many more. These ideas are not replaced or superseded by SCM. Rather, SCM incorporates all of these ideas to improve and manage the entire supply chain instead of just one firm in the supply chain.

Over the past 25 years, managers have learned to view their firms as a system of closely linked processes which deliver products and/or services to customers. Now managers are recognizing that their entire firm is just one link in a chain of firms whose purpose is to serve the customer. By increasing the integration of the entire supply chain, all the firms in the chain can increase their profits.

This book provides a basic introduction to help you understand SCM. It provides an introduction to the fundamental concepts in managing the flow of materials both inside an individual firm and throughout the supply chain. This book is organized in sections. The first section provides a brief history of supply chain management. The second section examines the fundamental business concepts such as operating environments, financial fundamentals, and an overview of the major managerial systems and tools used in SCM. The second section also examines customer linkages, including demand management. The third section provides an overview of the design and management of the transformation processes that will satisfy the demand. The final major section examines supply issues including inventory management principles, purchasing, and distribution.

The ability of a firm to engage in SCM places it onto the cutting edge for the 2000s and beyond. To gain an advantage through the use of SCM, each member of the chain will often have to prepare for some extremely significant changes in its business methods. To effectively improve service to the customer while reducing costs, each link in the chain must improve. The first step in doing this is to communicate the end customer's needs to all of the members of the supply chain. This requires a view of the entire system, which is often lacking. The second step is to have a management system in place which can communicate with the entire system and respond to information from the different components of the supply chain.

Information technology is a SCM tool. There is a surge of growth in interorganizational information systems, which means that firms are sharing information or at least data through their computer systems with other firms in their supply chain. For example, a junkyard can access a database that contains the records of inventory in stock at a large number of junkyards. This increases sales for all the junkyards involved while decreasing their inventory costs. It is information technology that allows this rapid communication across organizational boundaries. However, to be effective this information technology must be accompanied by information protocols and appropriate management practices. One consequence of information protocols is that the information technology may serve to create standards in a wide variety of market segments, which in turn may create a large market where only small isolated markets existed before.

The creation of standards for interfirm communication has long been a barrier to greater supply chain integration. Individual firms may have sought to have particular standards created which benefit them and consequently standard creation has slowed communication exchange. This has been a barrier to the wider use of electronic data interchange systems. Now, some technology is reducing this problem. For example, web-based interactions reduce the issue of noncompatible technologies being used by firms.

Acknowledgments

I owe thanks to many people for their help on this project and for their generous advice and time. My wife, Elaine Malinowski Fredendall, deserves the most thanks for her patience during the long weekends when I worked on this instead of helping around the house. My children, Susan and Joseph, also need to be recognized for their support.

I want to thank Ed Hill for involving me in this project and Drew Gierman who stuck with us during the long lead-time. The examples used in the book have frequently come from plants and facilities where I have toured or worked. The training to write my share of the book was given to me by many people, but a major portion of it was from my mentor, Dr. Steven A. Melnyk. I also want to thank my father-in-law, Wit Malinowski, who spent hours finding articles and references for me.

The Authors

Lawrence D. Fredendall is an Associate Professor at Clemson University, where he teaches and conducts research about supply chain management, the theory of constraint, and quality management. He earned his Ph.D. in operations management from Michigan State University in 1991, after having received his MBA with an emphasis in materials logistics management from Michigan State University in 1986. He received the "Jonah" certification by the Avraham Y. Goldratt Institute in 1992. He has published research articles in *Production and Inventory Management Journal, Journal of Operations Management, International Journal of Production Research, European Journal of Operational Research, Production and Operations Management, ATI: America's Textiles International, Journal of Managerial Issues, APICS — The Performance Advantage,* and *Organization Development Journal.* He was the editor of the Supply Chain Management Department for *APICS — The Performance Advantage* for the 2000 publishing year.

Ed Hill is a Principal with Chesapeake Consulting, Inc. of Severna Park, MD. He served over twenty years in the apparel and textile industry after completing his education in industrial engineering from Mississippi State University and serving his country as a First Lieutenant in the U.S. Army during the Vietnam era. His manufacturing experience has included positions as Vice President of Operations, Director of Manufacturing, Director of Engineering, and Plant Manager before becoming the Director of Clemson University's Apparel Research Center in 1987. In this role, he was responsible for a $10 million research project in which he led several studies to boost the competitiveness of the U.S. textile and apparel industry. In 1990, he participated in the landmark project led by Dr. Eliyahu Goldratt and John Covington to apply the theory of constraints concepts to synchronize the entire soft goods supply chain. As a full-time consultant since 1995, he has worked with numerous companies in a myriad of industries to apply synchronous flow and lean manufacturing techniques within supply chains of all sizes. He received the "Jonah" certification by the Avraham Y. Goldratt Institute in 1992.

About APICS

APICS, The Educational Society for Resource Management, is an international, not-for-profit organization offering a full range of programs and materials focusing on individual and organizational education, standards of excellence, and integrated resource management topics. These resources, developed under the direction of integrated resource management experts, are available at local, regional, and national levels. Since 1957, hundreds of thousands of professionals have relied on APICS as a source for educational products and services.

- **APICS Certification Programs** — APICS offers two internationally recognized certification programs, Certified in Production and Inventory Management (CPIM) and Certified in Integrated Resource Management (CIRM), known around the world as standards of professional competence in business and manufacturing.
- *APICS Educational Materials Catalog* — This catalog contains books, courseware, proceedings, reprints, training materials, and videos developed by industry experts and available to members at a discount.
- *APICS — The Performance Advantage* — This monthly, four-color magazine addresses the educational and resource management needs of manufacturing professionals.
- *APICS Business Outlook Index* — Designed to take economic analysis a step beyond current surveys, the index is a monthly manufacturing-based survey report based on confidential production, sales, and inventory data from APICS-related companies.
- **Chapters** — APICS' more than 270 chapters provide leadership, learning, and networking opportunities at the local level.

- **Educational Opportunities** — Held around the country, APICS' International Conference and Exhibition, workshops, and symposia offer you numerous opportunities to learn from your peers and management experts.
- **Employment Referral Program** — A cost-effective way to reach a targeted network of resource management professionals, this program pairs qualified job candidates with interested companies.
- **SIGs** — These member groups develop specialized educational programs and resources for seven specific industry and interest areas.
- **Web Site** — The APICS Web site at http://www.apics.org enables you to explore the wide range of information available on APICS' membership, certification, and educational offerings.
- **Member Services** — Members enjoy a dedicated inquiry service, insurance, a retirement plan, and more.

For more information on APICS programs, services, or membership, call APICS Customer Service at (800) 444-2742 or (703) 354-8851 or visit http://www.apics.org on the World Wide Web.

HISTORY AND INTRODUCTION

1

HISTORY
AND
INTRODUCTION

1 Introduction

The APICS dictionary defines the term *supply chain* as either the "processes from the initial raw materials to the ultimate consumption of the finished product linking across supplier-user companies," or as the "functions within and outside a company that enable the value chain to make products and provide services to the customer." The APICS dictionary defines *value chain* as those "functions within a company that add value to the products or services that the organization sells to customers and for which it receives payment."

Supply chain—1) The processes from the initial raw materials to the ultimate consumption of the finished product linking across supplier-user companies. 2) The functions within and outside a company that enable the value chain to make products and provide services to the customer.

APICS Dictionary, 8th edition, 1995

The differences between the definitions of the supply chain and the value chain are illustrated in Figure 1.1. In Figure 1.1 the supply chain is shown as a series of arrows moving from the raw materials stage to the final customer. Each of these arrows represents an individual firm, which has its own value chain. In Figure 1.1 this value chain is enlarged for one firm in the supply chain so that some of the internal functions of the firm that add value can be shown. In this example note that purchasing, marketing, and operations management are shown as part of the firm's internal value chain. These are internal functions of the firm and they occur in every firm that is a member of a supply chain.

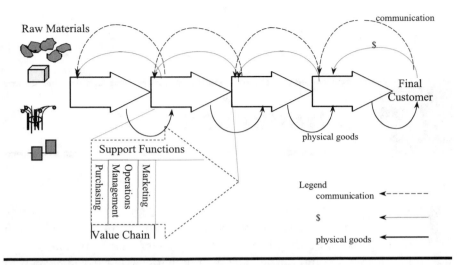

Figure 1.1   Supply Chain

Value chain—The functions within a company that add value to the products or services that the organization sells to customers and for which it receives payment.

APICS Dictionary, 8th edition, 1995

Another term used in some firms is *pipeline.* A pipeline is the supply chain for just one part used in a product. In these firms a supply chain for a complex product consists of many pipelines. An example of a pipeline would be a product that begins with rolled steel. A second step in the pipeline is the cutting process. This is followed by the stamping of the steel into a fender or other component. The component is then assembled into the final product. For example, it may be a fender which is assembled onto a car body.

Figure 1.1 also illustrates that the supply chain consists of more than the movement of physical goods between firms. It is also involves the flow of information between firms. This communication is necessary to manage and maintain the supply chain. Another supply chain flow is the flow of money. This is also shown in Figure 1.1 to illustrate that the primary purpose of every firm in the supply chain is to make money. This helps to remind all supply chain members that increasing their own income requires them to do everything in their power to improve the operations of the supply chain.

Supply Chain Management Evolution

Supply Chain Management (SCM) represents a significant change in how most organizations view themselves. Traditionally, firms view themselves as having customers and suppliers. Historically, a firm did not consider the potential for either its supplier or its customer to become a partner. In many industries each firm was very competitive with its suppliers and customers, fearing that they would be taken advantage of by them.

Beginning in the 1960s and 1970s firms began to view themselves as closely linked functions whose joint purpose was to serve their customers. This internal integration was often referred to as *materials logistics management* or *materials management*. In this structure those management functions involved in the material flow were grouped together. Firms that adopted the materials management structure integrated their purchasing, operations, and distribution functions to improve customer service while lowering their operating costs. Those firms that successfully integrated these functions did improve their performance. But, the firms were still constrained by other functions in the firm, which were not integrated, such as product development. Or, the firms were constrained by either their customer's or their supplier's unresponsiveness. These constraints prevented the firms from responding quickly to changes in the market which delayed their responses to meeting the changed needs of their customers.

Materials management—The grouping of management functions supporting the complete cycle of material flow, from the purchase and internal control of production materials to the planning and control of work in process to the warehousing, shipping, and distribution of the finished product.

APICS Dictionary, 8th edition, 1995

In the 1980s and 1990s many firms continued to further integrate their materials management functions. As it became clear that leading companies in this integration were able to increase their profits, more firms began to adopt supply chain management practices.

Power of Supply Chain Management

The power of supply chain management is its potential to include the customer as a partner in supplying the goods or services provided by a supply chain. Integrating the customer into the management of the supply chain has several advantages. First, integration improves the flow of information throughout the supply chain. Customer information is more than data. Customer information is data that has been analyzed in some manner so that there is insight into the needs of the customer. In the typical supply chain the further the members of a chain are from the end customer, the less understanding these members have of the needs of the customer. This increases the supply chain member's uncertainty and complicates the planning. Firms respond to uncertainty differently. Some firms may increase inventory while others may increase lead times. Either action reduces their ability to respond to their customers. As uncertainty is reduced because they have more information, firms are able to develop plans with shorter lead times. By improving the information flow in the supply chain, firms throughout the chain have less uncertainty to resolve during the planning process. This allows all the firms in the supply chain to reduce inventory and consequently to shorten their lead times while reducing their costs. This, in turn, allows the chain to respond to their customer faster.

A second advantage of integrating the customer into the supply chain is that this integrates the product development function with the other functions in the firm. This integration allows the product development staff to communicate more with the customer both internally and externally to the firm, which decreases the firm's response time to the customer's needs and tends to reduce product development time.

Some firms use the concept of internal customer to remind their employees that each employee performs just one step in a supply chain whose purpose it is to provide a good or service to the end customer. The purpose of the internal customer logic is to keep each employee focused on the needs of the end customer. This helps employees recognize that not only is their firm just one link of a larger supply chain, but that the firm itself can be viewed as a chain of processes each of which is a customer of the preceding process.

Internal customer—The recipient (person or department) of another person's or department's output (product, service or information) within an organization.

APICS Dictionary, 8th edition, 1995

By focusing on customers, particularly the end customer, all members of the supply chain see the need and benefits of obtaining and using information

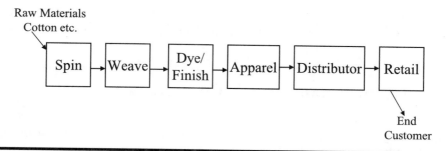

Raw Materials
Cotton etc.

Spin → Weave → Dye/Finish → Apparel → Distributor → Retail

End Customer

Figure 1.2   Example of Supply Chain

about the end customer. For example, Figure 1.2 is a simplified presentation of a supply chain for apparel sold at retail. If information from the end customer can be spread throughout the supply chain, there can be faster reaction from the supply chain to the end customer's requirements. If the retail firm shares its inventory status with the distributor, the distributor can prepare for reorders. If this information is shared with the apparel manufacturer, it can prepare for reorders also. As this information travels back through the chain and helps to eliminate surprises, the lead times for everyone can be reduced, which also reduces the amount of capital tied up in inventory. If enough information is being shared, the uncertainty in the demand faced by each step in the chain is reduced, which also leads to a reduction in inventory throughout the chain. Reducing uncertainty reduces the need for inventory in each level of the supply chain, because there is less need for just-in-case inventory. Since the supply chain members can use the information to produce inventory that is needed, the information sharing allows each firm to maintain or even improve its level of service. The following example illustrates the profit that can be made through proper supply chain management.

To illustrate, let's consider a supply chain with 3 members. In this hypothetical supply chain, if the finished goods inventory for Firm 3 (which sells to the retail store) can be reduced by 50 pieces, its raw materials can also be reduced by 50 pieces. Considering only the cost of its raw materials, which is $25, then the firm has reduced its capital invested by $1,250 (50 × $25) by using information to reduce inventory by 50 pieces. Assume that Firm 2 pays about $13 for the raw materials in the finished goods that it sells to its customer, Firm 3, at $25. If Firm 2 can reduce its finished goods and raw materials by the same 50 pieces Firm 3 was able to reduce, then the amount of its reduction in capital invested due to the elimination of the 50 pieces of inventory is $650 (50 × $13). For Firm 1, if it experiences the same reduction in inventory as the other firms, and its raw material costs are $6 for each part sold at $13, then its reduction in capital is $300 (50 × $6). The total savings in the supply chain due to the reduction of this inventory is $2,200 ($1,250 + $650 + $300).

What Is New about Supply Chain Management?

Supply chain management, just-in-time production (JIT), *quick response manufacturing, vendor management,* and other terms such as *agile manufacturing* all share the goal of improving vendor response to customer demand. All of these philosophies or concepts share the same core values. They attempt to improve customer service by eliminating waste from the system in all of its forms including wasted time. Supply chain management embraces the other philosophies and extends their scope from one firm to all the firms in a supply chain.

There are two forces driving supply chain management. First, is that there is the new communications technology available now that allows managers to actively manage a supply chain. Second, customers are demanding lower prices and better products and services. To meet their customers' demands, firms are optimizing the entire supply chain. Supply chain management allows all the firms in a supply chain to look beyond their own objectives to the objective of maximizing the final customer's satisfaction. The payoff for supply chain members that can do this is increased profits for their shareholders.

The largest barrier to successfully managing a supply chain is perhaps the human element. Failure to correctly manage the issues of trust and communication will abort any attempt to manage the supply chain. When there is a lack of trust and communication, the supply chain's members will soon succumb to greed or suspicion that other members of the supply chain are profiting at their expense. When the communication is not adequate, the supply chain will not improve its response enough to increase profits for its members. Without an increase in profits, the efforts to manage the supply chain will be reduced, because there will be no reward for actively managing it.

Supply chain management requires an unprecedented level of cooperation between the members of the supply chain. It requires an open sharing of information so that all members know they are receiving their full share of the profits. Since many of the firms in a supply chain do not have a history of cooperation, achieving the trust necessary for supply chain management is a time-intensive task.

Another way that the firms in the supply chain can save money is by ensuring that their marketing strategies correspond to the supply chain's capabilities—i.e., from their position in the supply chain they can actually provide what the customer wants. They are also able to gain money by improving the supply chain's capabilities to match the market demand with a decreased level of inventory. Firms are able to do this because they have additional information to forecast needs and as the lead time is reduced, their need to forecast is reduced. This reduced need to forecast reduces the need to carry inventory stocks for the just-in-case scenario.

Collaborative Planning

Many major retailers and large manufacturers have reduced their operating costs through their use of supply chain management techniques. But, there has been little effect on the price of the item to the consumer. Some argue that this occurred because the total amount of inventory in the supply chain was not reduced. Instead, the inventory may have been transferred to the second and third tier suppliers, but not eliminated from the supply chain.

Collaborative planning requires the firm to work with customers and suppliers to ensure that every day all of them have production and delivery schedules that agree with the needs of the customer. This has to be done routinely and not when the supply chain or a member is in a crisis.

Some firms are using Advanced Planning and Scheduling (APS) software to aid their collaborative communication. This is a trend that is just starting, but it may develop into a set of methods by which supply chain partners could have joint sales forecasts and/or production plans in which a revision by one partner would be immediately transmitted to the next partner.

For the APS to be effective for collaborative forecasting and planning throughout the supply chain, it is necessary that a reliable method for passing information between the different APS systems be in use. When a supplier has only one or two major customers, it is possible for them to have the same type of software as their customer. When a supplier has many customers, they cannot have software that matches each of their customer's needs.

How to Implement Supply Chain Management

A firm in the supply chain must initiate the attempt to form partnerships and actively manage the supply chain. Often a firm that has a large amount of market power in the chain will become the leader of the supply chain. This firm needs to justify the effort to manage the supply chain by explaining the benefits that will accrue to each member in the supply chain and to itself. To do this, the supply chain leader must show the partners where the improvements in the supply chain will arise and how these will lead to a gain for everyone. To establish trust among the members of the supply chain, the lead firm must also suggest how communication can be opened up and how every member will be ensured that it is receiving its fair share of profits. One recent example of this has been Wal-Mart. For years it has gathered extensive data on customer buying patterns. Wal-Mart has used this data internally to manage its own layouts and inventory. Now it is beginning to share all of this data with its most trusted suppliers. This will allow the supplier who knows how to take advantage of this data an opportunity to improve service to Wal-Mart while decreasing its own costs.

Managing a supply chain is more complex and difficult than managing an individual firm. But, the principles of management used to integrate a firm's own internal functions also apply to managing the entire supply chain. For example, a well-understood phenomenon in the management of a firm is that there is always a bottleneck that constrains sales. This bottleneck may be internal to the firm (a process that cannot produce enough to meet demand) or it may be external to the firm (market demand that is less than the capacity of the firm). This principle applies to the entire supply chain. While the supply chain is driven by customer demand, it is constrained by its own internal resources. One difference is that these resources may not be owned by the same firm. It is possible for the output of an entire supply chain to be limited because one firm does not have capacity to meet surging demand. It is also possible for every firm in the supply chain to be operating at a low utilization because there is not enough demand in the market for the products from the supply chain. There are bottlenecks inside the supply chain just as there are bottlenecks inside firms. To properly manage the supply chain, its members must be aware of the location of their bottlenecks internally and also of the bottlenecks in the supply chain.

Examining the Basics of Supply Chain Management

Since the principles of managing the supply chain are the same as those required to integrate the internal functions of the firm, this book will first explore the basic principles of supply chain management at an individual firm and then examine how these principles apply to the entire supply chain. There is no cookbook for managing the supply chain. However, there are some basic principles that can be used for it. And, a set of best management practices is evolving as managers gain experience managing their supply chains.

FUNDAMENTAL CONCEPTS

This second section provides basic information about how managers plan, organize, lead, and control. This information is used in the later sections when discussing how to manage the customer order cycle and how to manage efforts to improve the supply chain.

2 Management Basics

This chapter provides an overview of basic management practices. It then quickly reviews the basics of manufacturing strategy. It is important that supply chain managers understand how their decisions are shaped by and help to shape their firm's manufacturing strategy.

Many firms are now giving more and more responsibility for producing profit to those who actually do the production work. This means that it is increasingly important that everyone in a firm understand the basics of how to make money. This chapter begins with the concept of strategy and a short discussion of how strategy helps a firm to make money. Following this is a discussion of the performance measures, which are necessary to implement any strategy. These performance measures include financial measures that help a firm keep score or track whether it is making money. If business is a game, then we only know if we have won by the amount of money the firm made. Finally, the chapter covers the basic concepts and definitions used in many businesses.

The Manager's Role

Managers work in organizations, which are systems of people organized to perform some function. Managers are those people in the organization who direct the activities of other people in the organization. So, managers are people in an organization who get things done through and with other people. Managers are usually evaluated both on how efficient they are and how effective they are. Efficient managers are managers whose outputs meet or exceed the planned outputs or alternately the managers who achieve the most output for a given input. Effective managers are those who do the right things. The ideal manager is both effective and efficient.

Different lists of management functions have been prepared and presented over time beginning with Henri Fayol's list over 100 years ago, but they are all very similar. In this book the functions of a manager are to: (1) plan, (2) organize, (3) lead, (4) control, and (5) improve. The first basic function is to plan. This requires the manager to develop objectives, a strategy to meet the objectives, and plans to accomplish the goals necessary to implement the strategy.

Plan—A predetermined course of action over a specified period of time that represents a projected response to an anticipated environment to accomplish a specific set of adaptive objectives.

APICS Dictionary, 8th edition, 1995

Once they have a plan, managers perform their second function. They determine how to organize the firm. To do this, they must identify what tasks need to be done and who will do them. They then create a structure that is used for controlling the process later. Since managers do their jobs through people, they must lead these people. This requires the manager to understand how to motivate his/her employees, how to direct their activities, how to communicate effectively, and how to resolve conflicts as they arise. The fourth function of the manager's job is control. In the control function, managers monitor performance by comparing it to the established goals and then determining what actions must be taken to eliminate any significant deviations from the established goals. Managers exert control by using a control system. The control system is a method of collecting the necessary data and analyzing it so that appropriate action can be taken.

Control system—A system that has as its primary function the collection and analysis of feedback from a given set of functions for the purpose of controlling the functions. Control may be implemented by monitoring or systematically modifying parameters or policies used in those functions, or by preparing control reports that initiate useful action with respect to significant deviations and exceptions.

APICS Dictionary, 8th edition, 1995

The final element of the manager's job is to improve. This means that over time the firm must increase the productivity of each process. Productivity is measured as the ratio of process output to process input. For example, if a firm ordinarily produces 10 parts from 1 pound of steel in one hour, its productivity is increased if it can start making 11 parts from 1 pound of steel in one hour. However, the productivity measure must be used with care. Goldratt (1992) and others have pointed out the dangers of oversimplifying the productivity measure. For example, firms that pack their warehouses with unsold product may claim they are increasing their productivity because they are producing more, but because the only productive output is the output sold to the customer, the firm may have actually decreased its productivity. To prevent these costly errors and to help managers remain focused on the end customer, Goldratt suggests that managers only count product as output if it is sold.

Productivity—1) An overall measure of the ability to produce a good or a service. It is the actual output of production compared to the actual input of resources. Productivity is a relative measure across time or against common entities. In the production literature, attempts have been made to define total productivity where the effects of labor and capital are combined and divided into the output. One example is a ratio that is calculated by adding the standard hours of labor actually produced plus the standard machine hours actually produced in a given time period divided by the actual hours available for both labor and machines in the time period. 2) In economics, the ratio of output in terms of dollars of sales to an input such as direct labor in terms of the total wages. This is also called a partial productivity measure.
APICS Dictionary, 8th edition, 1995

Strategy

The most successful companies in any market have a strategic plan that is followed by all of its operating units. A strategic plan is the plan companies develop to meet their business goals. It is the mission that defines the business. The firm's top executives set the mission. It is this mission that determines the products or services that a firm will offer and consequently the mission determines who the customers of the firm are.

Strategic plan—The plan for how to marshal and determine the actions to support the mission, goals, and objectives. Generally includes an organization's explicit mission, goals, and objectives, and the specific actions needed to achieve those goals and objectives.
APICS Dictionary, 9th edition, 1998

Roadmapping is a similar concept to strategic planning. Firms create road maps by examining the firm in its environment. What are its products and markets? How are products developed, how is research and development conducted, etc.? The purpose of a road map is to determine where the firm needs to go. It is a detailed plan of the steps needed to move from where it is to where it wants to go.

Road maps can be created for entire industries as well as for companies. These road maps can identify areas needed for cooperation and coordination to improve conditions for the industry. This might include collaborative research programs. Or it may be that a different method of payment is needed.

Table 2.1   Levels of Strategy

Level	Decisions
Corporate Strategy	In what businesses should we compete?
Business Strategy	How should we compete in this business?
Manufacturing Strategy	How can we meet our business goals efficiently?

Typically strategy is broken up into three levels (see Table 2.1). At the top level is corporate strategy, which determines in what industries a firm will compete. For example, the decision about whether a firm will compete in the jet engine business and in television broadcasting (e.g., General Electric) is a corporate strategy decision. Once the corporate strategy decision is made, each strategic business unit (SBU) develops its own business plan for its products. The business strategy determines "HOW" a firm will compete in any given industry. For example, General Electric may compete on the basis of having the lowest cost in the appliance industry, but compete on the basis of having the best technology in the jet engine business. Creating the strategic plan for a business unit is the responsibility of the top executives in the business unit. They in turn submit their plan to the corporate office for approval. In a small company, which is only producing one product, the corporate and business strategy decisions are the same decision since the firm is in only one industry. The third level of strategy is the functional strategy or operations strategy. At this level, decisions about what a firm must do to fulfill its strategic business plan are made.

Manufacturing strategy—A collective pattern of decisions that act upon the formulation and deployment of manufacturing resources. To be most effective, the manufacturing strategy should act in support of the overall strategic direction of the business and provide for competitive advantages.

APICS Dictionary, 9th edition, 1998

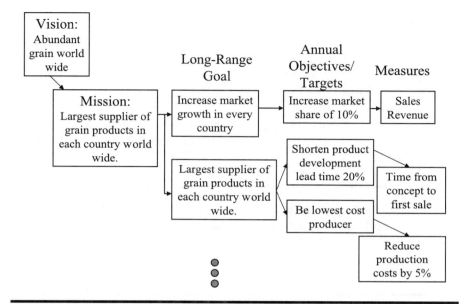

Figure 2.1   Strategic Plan Example

It is generally accepted that the crucial factors that determine whether one firm's strategy is successful while another firm's strategy fails are not necessarily the choice of the competitive priority (i.e., should the firm compete on the basis of low cost or high quality, etc.) but rather on how well the firm executes its chosen strategy. A major determinant of the firm's ability to execute any strategy is how well the strategy is understood throughout the firm. It is important that all members of the firm understand the strategy and their role in executing the strategy. A common framework to provide this understanding to everyone in the firm is illustrated in Figure 2.1.

In this example of communicating the strategic plan, the vision and mission are first explained. As the plan is communicated throughout the organization, specific measurable objectives are developed and communicated for each level. In this example both long-range goals and annual objectives are given. The long-range goal for the operations department is to be the lowest cost producer. Its immediate annual objective is to reduce processing costs by 5%.

The operations strategy is a collective pattern of decisions that are made at every level of the firm every day. These decisions are guided not only by the immediate annual objectives, but also by a manufacturing philosophy that helps guide all employees to make decisions that are compatible with the firm's objectives. If some individuals who are making these decisions do not understand the operations strategy, then their decisions are *not* likely to be aligned with the operations strategy. Some examples of well-known manufacturing philosophies are Just-in-Time (JIT), Theory of Constraints (TOC), and mass production. JIT focuses workers through its emphasis on eliminating waste in all of the various forms where it is found. TOC tries to focus everyone by emphasizing the need to identify the bottleneck and to do everything possible to enhance throughput from the system by managing the bottleneck. The mass production philosophy is to constantly emphasize the need to reduce cost.

Manufacturing philosophy—The set of guiding principles, driving forces, and ingrained attitudes that helps communicate goals, plans, and policies to all employees and that is reinforced through conscious and subconscious behavior within the manufacturing organization.

APICS Dictionary, 8th edition, 1995

Competitive Advantage

To maximize profits, firms seek to gain an advantage over their competitors in the market. This competitive advantage means that there is a reason why a customer would prefer to buy a product from one firm instead of from another firm. In some very tough markets, it may be necessary for a firm to gain a competitive advantage just to survive. When no firm has a competitive advantage over its competitors in the market, their products often become commodities. When a product is a commodity, the customers do not usually care which product they buy and consume, because from their point of view they can see no difference. For example, for many consumers, bread flour is a commodity. When most shoppers go to the grocery store they will select the cheapest bag of bread flour on the shelf. In this situation, it is difficult for a firm to make a lot of money. If one firm in the market cuts its prices then every firm must do so or lose all of its customers. Over time, in this type of industry, profits become smaller and smaller and these firms are constantly seeking ways to reduce their costs of production and distribution. To avoid having to sell their products as commodities, firms use marketing to establish a distinct brand name. Or they may try to distinguish their product in some other way. For example, one company stresses that it uses only the best wheat for its bread flour and that because of its distribution and operations methods, its flours are fresher than its competitors. It includes testimonials from pleased customers praising the distinct freshness and taste that the flour adds to their

- **Capacity** - amount, timing, type
- **Facilities** - size, location, specialization
- **Technology** - equipment, automation, linkages   Structural
- **Vertical Integration** - direction, extent, balance
- **Workforce** - skill level, wages, security
- **Quality** - defect prevention, monitoring, intervention
- **Production planning/materials control** -   Infrastructure
  sourcing policies, centralization, decision rules
- **Organization** - structure, control/reward systems, role of staff groups

Figure 2.2   Manufacturing Strategy Decision Categories (From Hayes and Wheelwright, *Restoring Our Competitive Edge:* **Competing through Manufacturing, 1984, p. 31. Reprinted by permission of John Wiley & Sons, Inc.)**

bread. This particular firm uses the "freshness" of its flour to justify charging a premium price for its flour.

> **Competitive advantage**—An edge, e.g., a process, patent, management philosophy, or distribution system, that a seller has that enables the seller to control a larger market share or profit than the seller would otherwise have.
>
> *APICS Dictionary,* 9th edition, 1998

The types of competitive advantages that a firm can obtain are usually broken into four categories. These are low cost, flexibility, delivery, and quality. A firm may have a competitive advantage in just one of these areas, or it may have a competitive advantage in all of these areas. Obviously a firm that is superior in all of these areas is a much tougher competitor in the market than one who has an advantage in only one area.

The ability of a firm to obtain a competitive advantage is determined by the decisions it makes about its structure and about its infrastructure. Hayes and Wheelwright (1984) grouped these decisions into eight categories (see Figure 2.2). A firm that makes these decisions in agreement with its strategic business plan creates manufacturing capabilities that allow the firm to compete in the way that was planned. Firms can change their direction, but to do so they must change each one of the manufacturing decisions made over the years. This is very difficult and takes a great deal of time and effort.

The first four of these eight decision categories are referred to as structural decisions. These are the decisions about capacity, facilities, technology, and vertical integration. These are the most difficult and expensive to change.

One decision about capacity is how much capacity to have. For example, should an automobile factory be designed to produce 50 cars an hour or 50 cars a day? Another issue with capacity is when to expand or contract capacity. For example, in a growing market a firm might project that overall demand will double every five years. Should it build the capacity to provide that demand before the demand emerges, or after the demand is actually present? Another capacity issue is the type of capacity a firm should have. Should it install flexible capacity that can be used for many products, or should it have capacity dedicated to one particular product. The advantage of flexible capacity is that a firm can respond to changes in the market without reinvesting. The advantage of dedicated capacity is that typically the variable cost per unit to produce is less, so that if there is sufficient volume the firm can be the low-cost producer.

Capacity—1) The capability of a system to perform its expected function. 2) The capability of a worker, machine, work center, plant, or organization to produce output per time period. Capacity required represents the system capability needed to make a given product mix (assuming technology, product specification, etc.). As a planning function, both capacity available and capacity required can be measured in the short term (capacity requirements plan), intermediate term (rough-cut capacity plan), and long term (resource requirements plan). Capacity control is the execution through the I/O control report of the short-term plan. Capacity can be classified as budgeted, dedicated, demonstrated, productive, protective, rated, safety, standing, or theoretical.

APICS Dictionary, 8th edition, 1995

The facilities decisions include the size of the facility, the location of the facility, and the extent to which the facility is specialized. In some types of firms the size of the facility is directly related to the capacity of the facility. For example, the size of a fast food restaurant will determine how many customers per hour can be served. In other industries, the size of the facility may allow for future capacity expansion. For example, a firm may establish a new facility and the building may be much larger than needed for present demand, because the firm wants a facility that can accommodate expected demand in the future. The location of the facility is important in many types of industry, but for different reasons. In manufacturing the location of the facility may be determined by the firm's need to access raw materials, to have a readily available supply of labor, or to have access to the customers or to other facilities of the firm. In services, a primary driver of location is usually access to the customer. A specialized facility is easier to manage than one that has many dissimilar tasks being performed. For example, it is often difficult to have high volume production in the same facility as a low volume production, because the nature of the management tasks are different. A firm can avoid this conflict by building smaller facilities that are specialized, or by building a large facility and managing the different products as if they were actually in different facilities. This plant-within-a-plant concept allows the firm to have more flexibility with its facilities.

Facilities—The physical plant and equipment.
APICS Dictionary, 8th edition, 1995

Technology refers to the collection of equipment, people, and procedures or systems that are used to produce the products or services the firm provides. A basic issue in every firm is the type of process technology it is using and whether this technology provides the capabilities that the firm needs to compete successfully in the marketplace. Technology management is an integrated effort. It cannot be done by just engineering alone, nor can it be done by operations alone. Technology management is concerned with the integration of the pieces into a system. Technology management integrates the resources and infrastructure to meet the goals and objectives of the firm. A fully integrated technology management program integrates all aspects of technology. This includes research and development, product design, manufacturing, marketing, distribution, human resources, purchasing, customers, and suppliers. Technology management is important for advanced supply chain management, but it is beyond the scope of this book, as we are addressing only the basics of supply chain management. In this book we will examine only those technology issues associated with the economics of the process technology, identifying the key problems associated with a process technology, the length of time to set up and use a given technology, and the flexibility and reliability of a given technology.

Technologies—The terms, concepts, philosophies, hardware, software, and other attributes used in a field, industrial sector, or business function.

APICS Dictionary, 9th edition, 1998

The last of the four structural variables is vertical integration. Vertical integration is concerned with determining the boundaries of the firm. What processes should be operated in house and what processes should be subcontracted to a supplier. The issues of sourcing are also a component of vertical integration. The decisions made about how to manage the supply chain belong to this decision category. For example, where should a firm buy its supplies and what types of relationships should the firm have with its suppliers? Where should the inventory stocks be held in the supply chain? Should every link in the chain hold its own inventory, or is it more economical to have one central location for storage of inventory?

Vertical Integration—The degree to which a firm has decided to directly produce multiple value adding stages from raw material to the sale of the product to the ultimate consumer. The more steps in the sequence, the greater the vertical integration. A manufacturer that decides to begin producing parts components, and materials that it normally purchases is said to be backward integrated. Likewise, a manufacturer that decides to take over distribution and perhaps sale to the ultimate consumer is said to be forward integrated.

APICS Dictionary, 8th edition, 1995

The second four of the eight decisions are referred to as infrastructure decisions. These include decisions about the workforce, quality, production planning/materials control, and the organization structure. Workforce decisions directly influence the capability of the firm. These decisions include the determination of the skill levels of its employees, and the wage policies, employment security offered by the firm, and training provided to the employees. The quality decisions include the type of quality system the firm should use and the firm's day-to-day operation of the quality system. Decisions about the production planning/materials control are concerned with the centralization of planning and decision rules used in the planning process. The final infrastructure decision concerns the basic organization decisions. Who should accept orders? How are due date commitments made? These include how the organization is structured, the details of its control/reward systems, and the role of staff groups in the firm. For example, should a firm use self-directed work teams? Should there be an incentive pool shared by everyone in the plant if goals are met?

These eight decisions create the system that everyone in the firm works within. By making these eight decisions in a coherent manner that focuses on providing the competitive priorities needed to fulfill the strategy, firms can achieve competitive advantages over their competitors. Before examining these concepts in greater detail, it is necessary to explore some other terms in greater depth.

Volume and Process Choice

To assist in making the facilities decisions, Hayes and Wheelwright (1984) tried to illustrate the decisions that need to be made about the size and design of a facility using the matrix in Figure 2.3. The matrix places the expected volume of the product on the top horizontal axis. The operations designer then moves down to the diagonal arrow, which suggests the type of layout that is suitable for the expected product volume. Given that type of layout, the designer can look across to the left vertical axis and see the resulting types of material flow that will exist. As technology improves there are more and more options that allow a greater variety of choice, but the choices of plant design are still constrained by the technology that is available for a given volume of production. One emerging technology is cellular manufacturing or group technology. Cellular manufacturing allows individual products of low volume to be combined into one larger family which may have high enough volume to be laid out on the shop floor in a way that creates a dominant product flow. This has the advantage of reducing material flow costs and simplifying control of the product flow through the plant.

It is easier to manage a shop that has dominant product flows. It is possible to have more automation of the production process and to spend less effort tracking the flow of materials when there are dominant product flows. However, the equipment to automate the production process and the material handling process is expensive and is cost effective only when there is enough volume.

Order-Winning Characteristic

Hill (1989) added market information to the ideas contained in Figure 2.3. Hill noticed that as companies changed over time, some companies were no

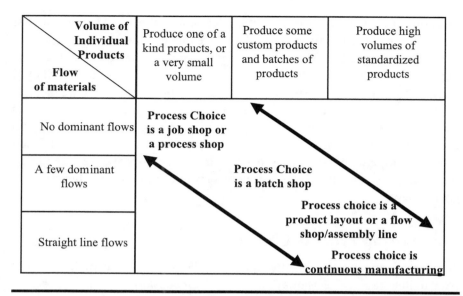

Volume of Individual Products / Flow of materials	Produce one of a kind products, or a very small volume	Produce some custom products and batches of products	Produce high volumes of standardized products
No dominant flows	**Process Choice is a job shop or a process shop**		
A few dominant flows		**Process Choice is a batch shop**	
Straight line flows			**Process choice is a product layout or a flow shop/assembly line** **Process choice is continuous manufacturing**

Figure 2.3 Volume and Process Choice (Adapted from Hayes and Wheelwright, Jan. 1979, pp. 133–140)

longer competitive. As he examined these firms he determined that as the markets of these firms changed, the firms had not changed their production process to match the needs of their new, emerging market. At most the firms would make small adjustments here or there. Further investigation about why companies would not adjust their production process to the market requirements revealed that managers at many companies did not recognize basic market changes such as changes in the order-winning characteristic (OWC). By order-winning characteristic, Hill means the characteristic of the product that causes the customer to select one particular product instead of another competing product. For example, while you are at the grocery store buying canned green beans, you will first examine the selection of green beans to decide which you want to buy. If you select the generic or store brand because of the low price, then price was the order-winning characteristic. Another customer may look at the same selection and choose a more expensive name brand of green beans because he/she likes the superior taste of the product. In this case the taste or quality of the green beans was the order-winning characteristic. Marketers might prefer to view this as buyers who are in different market segments.

What Hill established was that those companies that try to use one process to supply all the different segments of customers with the different versions of the product were not successful. Hill found out that with existing technology, some process choices were better at providing one order-winning characteristic than other process choices. For example, a production facility designed to produce high volume standard products is most capable of providing a product whose order-winning characteristic is price. It cannot easily provide customized products. While a job shop layout can provide a very customized product, it cannot provide the order-winning characteristic of price. Another important point about the concept of order-winning characteristics is that they change. Not only do different market segments have different order-winning characteristics, but the order-winning characteristics for one market segment this year may not be an order-winning characteristic for the market segment next year.

Order winners—Those competitive characteristics that cause a firm's customers to choose that firm's goods and services over those of its competitors. Order winners can be considered to be competitive advantages for the firm. Order winners usually focus on one (rarely more than two) of the following strategic initiatives: price/cost, quality, delivery speed, delivery reliability, product design, flexibility, after-market service, and image.

APICS Dictionary, 9th edition, 1998

Order-Qualifying Characteristic

Hill also described order-qualifying characteristics (OQC). OQCs are the product characteristics that customers demand, but they are not characteristics that customers see as distinctive enough that they will buy the product because of those characteristics. The customer just expects these OQCs to exist in a product before they will even consider purchasing it. For example, when many customers go to buy a new automobile they decide on the price range they are willing to pay (say, $20,000 to $25,000) before they begin to investigate automobiles. When they start searching for automobiles they will consider only autos that are within this price range. That is, the only autos that are qualified for future consideration are priced below $25,000. So, price is the OQC in this example. The OQCs change due to competitive pressure and the needs of the customer. For several years quality may be the order-winning characteristic, but as every member of the industry starts to produce products with roughly comparable quality, the quality of the product could rapidly become an order-qualifying characteristic and not an order-winning characteristic. As that happens a OWC such as product design may emerge.

Order qualifiers—Those competitive characteristics that a firm must exhibit to be a viable competitor in the marketplace. For example, a firm may seek to compete on characteristics other than price, but in order to "qualify" to compete, its costs and the related price must be within a certain range to be considered by its customers.

APICS Dictionary, 9th edition, 1998

Order-Losing Characteristic

Another related concept is the idea of an order-losing characteristic. An order-losing characteristic is a characteristic that when present will cause potential customers to refuse to consider the product. Lack of safety is often an order-losing characteristic. If there is publicity about catastrophic failures of a product, many customers will not consider purchasing the product. So, the lack of perceived safety is an order-losing characteristic.

Product Life Cycle

One characteristic that makes it difficult to design any production process is that all products go through a product life cycle as shown in Figure 2.4. When a product is first introduced it has low volume. For example, when the personal computer first came on the market there was demand for only a few at a time. In this "introductory" stage the volume was low enough that the suitable production method was similar to a job shop. As the product improved, demand for the product grew and entered the "growth" stage of the cycle. At this stage of the cycle the demand was large enough for batch production. As the market increased, IBM introduced its PC and produced it using an assembly line. This PC standardized the product offering and the sales volume increased sufficiently to support an assembly line. To survive, IBM's competitors had to change their production methods and product offerings to at least match IBM. During each stage of this product life cycle, companies were always faced with two choices: change the production system to one that can cost effectively provide customers with the product that meets the market requirements (i.e., the product has the desired order-winning characteristic) or go out of business.

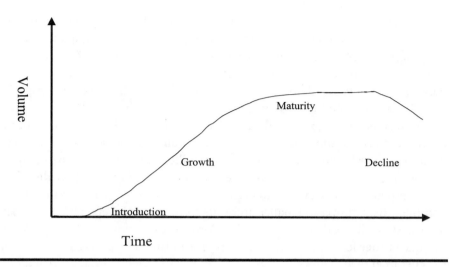

Figure 2.4   Product Life Cycle

Four Competitive Priorities

As discussed earlier in this chapter, the basic approaches that companies take to compete with each other in a market segment are grouped into one of four categories or competitive priorities. These are cost, quality, delivery, and flexibility.

To compete on the basis of cost, a firm tries to be the lowest cost producer of a given product for a specific market. This gives the firm the ability to underprice its competitors if it needs to do this. For example, you work for a brewery that is able to produce and distribute beer cheaper than any other brewery in the market. You could sell it at the same price as the beer of the other breweries in the market and make more money than the other breweries. In addition, this low cost production allows you to win any price war that your competition may start because you can make money on your beer at lower prices than they can.

To compete on the basis of quality, a firm determines what quality characteristics its customers desire and attempts to be the best at providing those to the customer. For example, if you are selling cars and a major quality characteristic that customers value is that the car be defect-free for the first 60 days, your firm would need to improve the process and design to provide that to the customer better than your competitors can provide it.

To compete on the basis of delivery, a firm concentrates on providing shorter and more reliable lead times than its competition. This is often the case with a make-to-order firm. A make-to-order firm does not carry product in stock, but manufactures it after it receives the order. If your firm manufactures propellers for planes, you may have a competitive advantage if you can manufacture a new propeller to replace a damaged propeller faster than your competition can. Since the damaged plane cannot fly until the damaged propeller is replaced, the customer may even pay a premium price for the faster delivery.

To compete on the basis of flexibility, a firm develops the ability to respond quickly to the customer needs. This flexibility can exist in many different areas if the customer is willing to pay for the flexibility. For example, a firm may be flexible in the quantity that a customer can order from them. Many distributors compete in this way, by buying in large quantities from a manufacturer and then distributing in smaller amounts to their customers. Or a firm may be flexible in its lead times to produce orders for customers. It may have a standard order lead time and a rush order lead time if it is an emergency for the customer.

Finally, many firms are able to compete by building on their capabilities in one priority area to develop unique capabilities in another. For example, as a firm produces better and better quality it develops capabilities in its process design, product design, and manufacturing abilities, which can lower its costs. This allows the firm to compete in both areas simultaneously.

Value

Many customers do not consciously evaluate whether one product has this particular competitive priority or that competitive priority. Customers do evaluate the value a product has for them. They may do this consciously or unconsciously, but they are making a value choice when they purchase a product. A very important point about value is that it is the customer's perception that determines the value of the product. Melnyk and Denzler (1996) describe this using the following "Value Equation."

$$\text{Value} = f(\text{quality, delivery, and flexibility})/\text{cost}$$

This description of value states that value is a function of quality, as perceived by the customer; the delivery of the product which could represent the actual lead time, variance in the lead time, or the method of delivery; and the firm's flexibility. Further, the equation states that customers perceive these benefits of the product in relation to the cost of the product.

Notice that the value equation contains the competitive priorities and that these are also the order-winning and order-qualifying characteristics. To illustrate this, think of a customer examining a product. If the product does not have an order-qualifying characteristic that the customer wants, the customer will perceive that the product has a low value. For example, if I want to buy an ice cream cone on a hot Saturday afternoon, I may decide that I want a hand dipped cone. Because I am taking all of my kids with me, I want to go where we can choose from a variety of flavors. I am willing to eat it there, but I want it served immediately and I want someone who can mix and match toppings. But, I do not want to spend more than $10 for the four of us. I may choose not to go to the yogurt place because of the limited selection of flavors. Further, I do not consider the small dairy store, a few blocks over, because it does not have a large enough selection of toppings and they are not flexible. For example, they will only put toppings onto the ice cream in a portion size that they find adequate. In this example, I am obviously putting a high value on flexibility (i.e., flavor toppings variety) and on price. These would be order-winning characteristics for my decision this time. An order-qualifying characteristic is that the cone be hand dipped. An order-losing characteristic (e.g., the small dairy store) is a limited selection of flavors. My value equation (as shown below) would place great emphasis on quality and flexibility and medium interest on the cost. Delivery was classified as an order-winning characteristic since I am considering only stores within a predetermined driving range. In this value equation the order-winning characteristics are shown in bold to demonstrate that these are the important characteristics.

$$\text{Value} = f(\textbf{quality} \ (OQC), \text{delivery}(OQC) \text{ and } \textbf{flexibility}(OWC))/\textbf{cost}(OWC)$$

Design for Manufacturability

Design for Manufacturability (DFM) is based on the premise that product designers can develop a product design at the same time that the manufacturing engineers are developing a process to manufacture the product. This allows both groups to more accurately track product cost and to ensure that the final product can be produced. This requires teamwork between the two groups of engineers.

This is a change from the sequential design process where the product design engineers develop a design, then throw it over the wall to the process engineers to develop a manufacturing process. The serial process creates added time in the product design process because the design may be tossed back and forth between the two groups for a while. There is also a lack of accountability for the final product, with each group blaming the other group for final decisions.

Some argue that up to 75% of a product's manufacturing costs are determined when it is designed. A design that is manufacturable does not require later tweaking.

Many companies are now teaming up with suppliers to incorporate DFM into products produced by suppliers. The first step in this is to have the suppliers' engineers review designs during the development stage to help solve any engineering dilemmas that occur.

The basic principle behind DFM is to reduce the number of parts and to obtain simpler designs. Products with modular construction are cheaper to assemble. DFM also focuses on adding functions to the product and on increasing quality and reliability. Both of these are aided by increased simplification of the product design.

DFM is usually implemented by a multifunctional design team. A core team is established for each project, which might consist of the project engineer, a designer, representatives from marketing, technical support, purchasing, factory line assemblers, and foremen. The entire team meets regularly and works together on the project after they receive training in the principles.

It is important that these teams be empowered to work across functional boundaries. The team must meet regularly and be required to review design and assembly alternatives. The team must also physically run through the fabrication assembly on an ongoing basis to optimize the product.

It is possible to include the customer of the product on the design team. This will allow input form the customer to the final design. The team needs to use the principles shown in Table 2.2.

Table 2.2   DFM Axioms

Minimize the number of parts	Minimize part handling and presentation
Minimize the number of part variation	Avoid flexible parts
Use modular design	Design parts to self-fixture
Avoid fasteners	Maximize part symmetry or asymmetry
Use multifunctional parts	Maximize visibility of assembled area
Design for top down assembly	Avoid hazardous material or processes
Maximize part mating or compliance	

Adapted from Ingalls, 1996.

3 | Performance Measures

eveloping and using performance measures is an essential function of management. Managers give direction and achieve control through the use of performance measures. Companies are in business to make money. To know if they are making money they measure the amount of profit that they make from the sales of their products. Companies measure profit in many different but related ways, depending on the decision that they are trying to make.

Profit—1) Gross profit—earnings from an on-going business after direct costs of goods sold have been deducted from sales revenue for a given period. 2) Net profit—earnings or income after subtracting miscellaneous income and expenses (patent royalties, interest, capital gains) and tax from operating profit. 3) Operating profit—earnings or income after all expenses (selling, administrative, depreciation) have been deducted from gross profit.

APICS Dictionary, 8th edition, 1995

Net Profit

The net profit or net income of a business is the amount of money left over after variable costs (i.e., costs which are directly related to the amount of product being produced) and fixed costs (i.e., those costs which must be paid even if the firm produces nothing) are subtracted from the total sales revenue. For example, if a company has total sales of $1,000,000, fixed costs for the plant and all salaried personnel of $500,000, and variable costs for selling expenses, material, transportation, etc. of $300,000, then their net profit is $200,000. This is calculated as: $1,000,000 − $500,000 − $300,000 = $200,000 before income taxes. If the income tax rate is 10%, then their taxes are $20,000 (10% × $200,000) so the net profit after taxes is $180,000.

Break-Even Point

Often to analyze the impact of decisions, companies calculate the break-even point to produce a product. The break-even point is the production level where the volume of sales pays both the fixed costs and the variable costs of producing that product. If the total revenue and total costs were graphed, the break-even point is where these two lines intersect. Of course this intersection is where the number of sales (i.e., total sales revenue) equals the total expenses.

Break-even point—The level of production or the volume of sales at which operations are neither profitable nor unprofitable. The break-even point is the intersection of the total revenue and total cost curves.
APICS Dictionary, 8th edition, 1995

This is illustrated in Figure 3.1, which also shows the fixed costs as a horizontal line. Remember, the fixed costs (FC) are horizontal by definition, since they are always the same. In this example, the fixed costs are $50,000, so even if the company does not produce any products, its fixed costs are still $50,000. The variable costs (VC) are 0 when no product is produced and increase by the same amount for each unit produced. So, if the variable costs are $1 for each unit, then producing 100 units increases variable costs by $100.

The break-even point is calculated by setting the total cost (TC) equation equal to the total revenue (TR) equation:

$$TC = FC + VC \times volume = volume \times price = TR$$

Rearranging this equation to find the number of units or volume that need to be sold to break even we have:

$$Volume_{break\text{-}even} = FC/(Price - VC)$$

A common example of this occurs daily at the famous widget company. Assume that you work for the famous widget company and everyone is trying to buy your widgets. Just to be sure that you are making money on them at the current price you decide to calculate the break-even point for your widgets. To

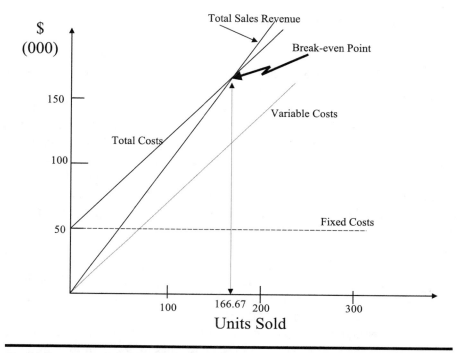

Figure 3.1   Break-Even Point Example

do this you gather the following data. The sales price is $1,000. The unit variable costs or the variable costs to produce and sell one widget are $700. Finally, the fixed costs for the facilities to produce the widget and management salaries to produce the widget are $50,000. So, you calculate the break-even point as:

$$\text{Volume}_{break-even} = FC/(Price + VC) =$$
$$\$50,000/(\$1,000 - \$700) = 166.67 \text{ widgets}$$

To make it easier to understand what this means and to present the information to your fellow managers, you draw the break-even chart shown in Figure 3.1. The break-even chart shows the point of intersection of the total revenue line and the total expenses line as approximately 167 widgets.

When creating this break-even chart you first create the horizontal and vertical axes. The volume of unit sales always goes on the horizontal or x axis. The revenue or dollars always go on the vertical or y axis. Next you graph the fixed costs onto the chart. The fixed costs are a straight line, which is always parallel to the horizontal or x axis. In Figure 3.1 the fixed costs are $50,000 and are shown as a dashed line. Once the fixed costs are graphed, you then graph the variable costs. The variable costs line will always start at 0, since there will be no variable costs if you do not produce any units. You need to calculate another point for the variable costs. It does not matter what point you pick as long as it is on the *x* axis. In Figure 3.1 the variable costs were calculated for 100 units of sale. When 100 units are sold the variable costs are $70,000 (i.e., 100 × $700). The variable cost line is then drawn from the origin through the point (100, $70,000). This is shown in Figure 3.1 as a dotted line. The total cost line is the sum of the fixed cost line and the variable cost line. The total cost line will always start at the fixed cost for 0 volume and it will be parallel to the variable cost line. To find a second point to draw the variable cost line, just add the fixed cost to the variable cost you calculated above for the sale of 100 units. In this example the total cost line starts at the point (0, $50,000) and goes through the point (100, $120,000), since the total cost when a 100 units are produced and sold is $50,000 fixed cost + $70,000 variable cost. The total cost is shown in Figure 3.1 as a solid line. Finally, the total revenue curve is graphed. This starts at the origin since there is no revenue if there are no sales. The same number of unit sales used to calculate the variable cost line can be used to graph the total revenue line. In this example at 100 units of sale the total revenue is $100,000. The total revenue line is shown as a solid line in Figure 3.1. The break-even point is where the total revenue and total cost lines intersect and the number of units at this point can be found by drawing a line from this intersection down to the horizontal axis as shown in Figure 3.1 as the dotted line with two arrow heads.

The implications of the break-even point are significant. It is obvious from Figure 3.1 that a lower break-even point means more profit for a given sales volume. A lower break-even point also means that a company can make money at a lower sales volume. From the break-even equation you can see that there are two ways to reduce the break-even point. The first is to increase the contribution margin (i.e., the difference between the sales price and the variable costs). The meaning and the influence of the contribution margin is discussed below. Obviously the second way to lower the break-even point is to reduce the fixed costs, since fewer units need to be sold to pay these bills. This is necessarily a simplified discussion of fixed and variable costs. In the long run, no costs are truly fixed, and many costs are semi-fixed and semi-variable (i.e., they behave as fixed costs within a certain volume range.)

In the example above the contribution margin is $1,000 sales price minus the $700 variable cost or $300. This is the denominator in the break-even point calculation. This is shown in the equation below:

$$\text{Volume}_{\text{break-even}} = FC/(\text{contribution margin})$$

This equation makes it clear that we can make more money by either reducing the fixed costs or increasing the contribution margin, because either approach will reduce the break-even point for us.

Contribution margin—An amount equal to the difference between sales revenue and variable costs.

APICS Dictionary, 9th edition, 1998

Cash Flow

It is important for firms to project their cash flow. This ensures that they will have funds on hand to pay their bills and their payroll and to invest in new projects. Cash flow is watched carefully by the finance department, because it is possible for a firm to make a profit on paper and still go broke because they do not have the cash needed to pay bills. When a firm invests in a new plant or new equipment, someone estimates the cash flow to calculate how the investment will influence the firm's ability to pay bills and dividends. In the example above the expenses were shown only as fixed or variable and the calculation was not concerned with the time period in which the expenses had to be paid. Likewise, the sales revenue was calculated on a per unit basis, but the sales may come at different time periods and some of the money for the sales may not be collected immediately. Based on historical records and contracts with suppliers and customers, the firm estimates its cash flow over the relevant time period. The cash flow analysis is shown in Table 3.1 for the example above. The cash flow analysis provides new information that the break-even analysis did not provide. It uses the period-by-period forecast to estimate when the firm will receive money and when it will pay out money. For the first seven months of the project the firm will have negative cash flow of $56,000. Then as sales pick up it will alternate positive and negative cash flows. Part of the reason for this is that revenue will be received 30 days after the product is sold, but the expenses to produce the product will be paid the month earlier.

Cash flow—The net flow of dollars into or out of the proposed project. The algebraic sum, in any time period, of all cash receipts, expenses, and investments.

APICS Dictionary, 9th edition, 1998

Table 3.1 Example of Cash Flow Analysis

						Month						
	1	2	3	4	5	6	7	8	9	10	11	12
Units Sold							10	10	20	20	30	30
Sales Receipts								10000	10000	20000	20000	30000
Sales Expenses								1000	1000	2000	2000	3000
Material Costs							5000	5000	10000	10000	15000	15000
Other Unit Costs								1000	1000	2000	2000	3000	3000
Equipment	10000			10000		10000						
Salaries					10000	10000						
Total Expenses	10000	0	0	10000	10000	20000	6000	7000	13000	14000	20000	21000
Total Revenue	0	0	0	0	0	0	0	10000	10000	20000	20000	30000
Cash Flow	−10000	0	0	−10000	−10000	−20000	−6000	3000	−3000	6000	0	9000

The cash flow analysis in the example above neglects one very important point. It is important to recognize that $1 today is worth more than $1 tomorrow. With the $1 received today there is the potential of investing it so that it earns money. This concept is called *present value*, which is defined by the *APICS Dictionary* as "the value today of future cash flows." The net present value considers the value today of "future earnings (operating expenses have been deducted from net operating revenues) for a given number of time periods."

Present value—The value today of future cash flows. For example, the promise of $10 a year from now is worth something less than $10 in hand today.

APICS Dictionary, 8th edition, 1995

Net present value—The present value of future earnings (operating expenses have been deducted from net operating revenues) for a given number of time periods.

APICS Dictionary, 8th edition, 1995

To calculate the net present value (NPV) of an investment, a firm's accounting department estimates the firm's minimum desired rate of return. This is used to discount the cash flows above. For the cash flow example above, if we assume that the minimum desired rate of return for that firm is 10%, then the cash flow for each period is multiplied by the present value of a dollar for that period (i.e., $P = S/(1 + r)^n$ where S = $1, r is the interest rate or the desired rate of return, and n is the number of periods in the future) to find the present value of that cash flow. This is shown in Table 3.2.

Table 3.2 NPV of Cash Flows

	Present Value of $1 at 10% for Each Month											
	1	2	3	4	5	6	7	8	9	10	11	12
	1	.909	.826	.751	.683	.621	.564	.513	.467	.424	.386	.350
Cash Flow	−10000	0	0	−10000	−10000	−20000	−6000	3000	−3000	6000	0	9000
Present Value	−10000	0	0	−7510	−6830	−12420	−3384	1539	−1401	2544	0	3150

The values for the row labeled cash flow in Table 3.2 came from the row labeled cash flow in Table 3.1. An example of the calculation for the present value of $1 row of Table 3.2 is shown below (note, that the present value of cash received in the present or in this case month 1 is not discounted since we can spend it now). Because month 2 is one period in the future $n = 2 - 1 = 1$

$$P_2 = S/(1 + r)^n = \$1/(1 + .1)^1 = .909$$

This present value is multiplied by the cash flow for the period to determine the present value of that cash flow. In month 4, the present value of $1 is 0.751. The cash flow is −$10,000, which means that we are investing $10,000. The present value of this $10,000 investment is found by multiplying 0.751 by $10,000.

The net present value of the investment is the sum of each period's present value. In this example the NPV is negative (i.e., −34,312). So, at an interest rate of 10%, this project is not worth investing in. Given this information it is necessary to investigate whether the cash flows can improve dramatically. For example, how could managers increase revenue earlier? Could the equipment start operating faster? Could sales be increased or could costs be decreased?

Return on Investment (ROI)

A basic measure of how well a firm is meeting its goal of making money is the return on investment (ROI) that the firm is achieving. The ROI measurement tells us how good the net profit we achieved is compared with the performance of others. It is calculated as:

$$\text{ROI} = \frac{\text{Net Profit}}{\text{Invested Capital}} \times 100$$

ROI is a performance measure. It is a method that the owners of the company use to evaluate how well the capital they have invested in a company is performing. It allows them to compare their investment in this company to investments elsewhere. It can be divided into its component parts for a more thorough analysis, but in this case we will consider only the basic case.

Return on investment (ROI)—A financial measure of the relative return from an investment, usually expressed as a percentage of earnings produced by an asset to the amount invested in the asset.

APICS Dictionary, 9th edition, 1998

If the net profit of a firm is $100,000 for a year, we do not know how good this is compared to other investments. If the invested capital in the firm is $1,000,000 then this is an ROI of 10%. If the invested capital is $10,000,000 then the ROI is 1%. This is calculated as:

$$\text{ROI} = \frac{\$100,000}{\$10,000,000} \times 100 = 1\%$$

Internal Measurements

The measurements of net profit, cash flow, and ROI tell the manager whether a firm is making money and what its relative performance is. Obviously when managers need to make decisions in the course of daily business, they will try to increase profit, cash flow, and ROI. But, it is not always immediately obvious in many decisions which alternative will maximize this return. The break-even analysis and present value analysis do provide information about investments, but they do not provide insight into decisions about how to schedule the equipment and which orders to accept given certain capacity constraints.

To make internal operating decisions all firms have developed internal measurements. The purpose of these internal measurements is to guide all of the firm's employees to make the decisions that will provide a firm with the highest possible ROI, net profit, and cash flow both now and in the future.

Throughput

One internal measurement used by some firms is called *throughput*. When used in the theory of constraints as developed by Goldratt (1992), throughput is the money generated by goods sold. In practice, throughput is calculated as the net sales minus the material cost of the goods sold. If the net sales were $1,000,000 and the cost of goods sold was $700,000, then the throughput is $300,000.

Throughput = Sales Revenue − Material costs of goods sold

or

$1,000,000 − $700,000 = $300,000

Throughput is used as an internal performance measure because it links the local action of scheduling a machine or selecting an order to run next on the machine to the net profit of the firm. Throughput considers both the sale of the finished products and the purchase of raw materials and components to produce the product. Throughput is actually the contribution margin of the product if the only variable cost is the cost of the raw material.

Throughput (T)—In the theory of constraints, the rate at which the system (firm) generates money through sales. Throughput is a separate concept from output.

APICS Dictionary, 9th edition, 1998

Inventory

There are two different methods of measuring inventory. They differ on whether value added is included when calculating the value of the inventory.

Inventory—1) Those stocks or items used to support production (raw materials and work-in-process items), supporting activities (maintenance, repair, and operating supplies), and customer service (finished goods and spare parts). Demand for inventory may be dependent or independent. Inventory functions are anticipation, hedge, cycle (lot size), fluctuation (safety, buffer, or reserve), transportation (pipeline), and service parts. 2) In the theory of constraints, inventory is defined as those items purchased for resale and includes finished goods, work in process and raw materials. Inventory is always valued at purchase price and includes no value-added costs, as opposed to the traditional cost accounting practice of adding direct labor and allocating overhead as work in process progresses through the production process.

APICS Dictionary, 9th edition, 1998

Inventory is often used as an internal performance measure, for example, in just-in-time (JIT) production, because inventory measures how effectively waste has been reduced in the process. In a JIT system, the level or amount of inventory serves as a thermometer of the waste in the system. If quality is improved, then a firm does not need to carry as much inventory, because it does not need the buffer or reserve to protect its production process from a batch of poor quality parts. If maintenance improves, it is likely that machines will not break down as frequently and the inventory buffer or safety stock can be reduced. If the process design improves, then less inventory is needed to protect the process from problems. If the product design is improved, less inventory is needed because there may be fewer parts, or because it is easier to manufacture the product and less inventory is needed as protection. If employee motivation improves, then there is less need for inventory because the employees will respond faster to problems and interruptions. So if a firm were able to reduce its inventory level from enough inventory for 2 weeks operations to enough inventory for 1 week of operations, it would believe that it had improved. This is often shown as in Figure 3.2 as a ship floating on the sea. If the sea level drops, then the ship hits a rock. In a factory the inventory is the sea. If the process keeps operating as we drop the inventory level then we have successfully eliminated all of the rocks that could sink the ship. On the side of Figure 3.2 is a thermometer to illustrate that inventory is only a measure of the system's performance just as the thermometer is only a measure of the temperature.

In the theory of constraints (TOC) system, it is important to value inventory at the price for which the material was purchased. So, finished goods are carried in inventory only at the cost of the raw materials in them. The cost of

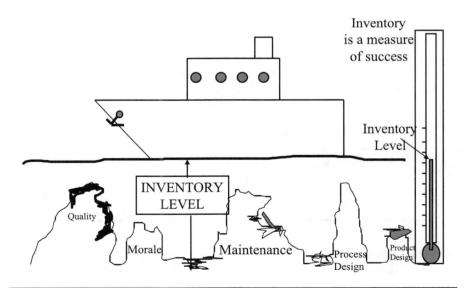

Figure 3.2   The Process Floats on a Sea of Inventory

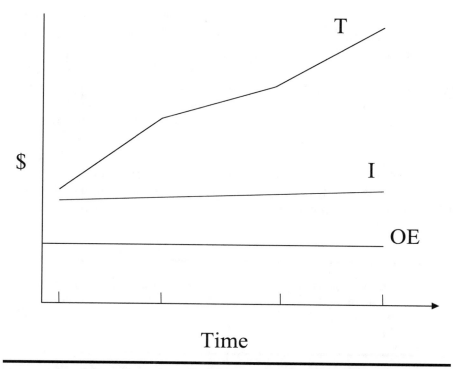

Time

Figure 3.3 Tracking Local Performance Measures over Time—Throughput (T), Inventory (I), and Operating Expenses (OE)

the labor to produce them is not included, since including the concept of value added in the finished goods inventory would allow the inventory measure to be manipulated to show increased profits when products are actually in warehouses. In the TOC system, inventory is a system performance measure. In TOC, if inventory can be reduced while throughput is increased then it is a measure of the firm's success. This is illustrated in Figure 3.3.

Operating Expenses

The two local measures described so far have been throughput or the sales of the product and inventory or the material held to produce the product. It is important also to have some type of measure of the expense of manufacturing the product. In TOC, this measure is called *operating expenses.* Operating expenses are the monies spent converting raw materials into sales in a specific time period. In the TOC system all overhead costs, depreciation, administrative expenses, labor (both direct and indirect), marketing expenses, utility costs, etc. are lumped together and called operating expenses. The measure of operating expenses is not shown on the balance sheet and income statements given to the shareholders. Instead, the use of operating expenses is a local performance measure. It is designed to give the line workers or line supervisors a means of focusing their decisions so that they can help increase the net profit of the firm. For example, a decision to increase the overtime budget of a department could be viewed as a decision about whether to increase the operating expenses of the firm. The determining factor in the decision would be whether the firm's throughput with that overtime would increase faster than its operating expenses.

The role of the local performance measures is to guide the many internal decisions made in daily operations, so that the firm's throughput goes up at a faster rate than inventory and operating expenses. This increases the net profit.

Operating expenses—In theory of constraints, the quantity of money spent by the firm to convert inventory into sales in a specific time period.

APICS Dictionary, 8th edition, 1995

The shop floor measures of throughput, inventory, and operating expenses can be translated into profit, return on investment, and cash flow and vice versa. Consider an example where Throughput (T) equals $1,000 and Operating Expenses (OE), Cost of Materials, and Inventory are equal to $500, $1,000, and $2,000, respectively. Since T = Sales Revenue − Raw Materials Cost, the Sales Revenue for the period being considered is $2,000 (i.e., Sales Revenue = T + Raw Materials Cost). The profit is T − OE, so the profit is $1,000 − $500 = $500. The Return on Investment (ROI) is $500/$2,000 which equals 25%. (Remember, the inventory measure is the value of all non-depreciated assets including material).

Standard Times and Efficiency

Standards are important elements of managing a business. Standard times and efficiency measures are internal performance measures. The standards establish the norms and allow a firm to measure and control its performance. The standard time allowed to complete a job or to complete a certain amount of work is one of the most common internal performance measures in manufacturing.

The standard time is used in determining machine and labor requirements. The standard time is calculated as the time that an average worker, following prescribed methods and taking an allowed amount of time for rest, takes to perform a task. It can be used to calculate incentive pay systems and to allocate overhead to products. It is sometimes called *standard hours.*

Standard—1) An established norm against which measurements are compared. 2) An established norm of productivity defined in terms of units of output per set time (units/hour) or in standard time (minutes per unit). 3) The time allowed to perform a specific job including quantity of work to be produced.

APICS Dictionary, 9th edition, 1998

Standard time—The length of time that should be required to (1) set up a given machine or operation and (2) run one batch or one or more parts, assemblies, or end product through that operation. This time is used in determining machine requirements and labor requirements. Standard time assumes an average worker following prescribed methods and allows time for personal rest to overcome fatigue and unavoidable delays. It is also frequently used as a basis for incentive pay systems and as a basis of allocating overhead in cost accounting systems.

APICS Dictionary, 9th edition, 1998

It requires a lot of effort to accurately measure the standard times to complete any procedure. Because the standard times are used to evaluate the worker and sometimes to create pay standards, it is important that the standard times be evaluated as the work methods change. The standard time depends on the work method in use. The work method includes the use of the tools and facilities used to complete the task as well as the operation motions, the station layout, the position of the material, the types of materials used, and the working conditions. In many companies it is the responsibility of the time study department or industrial engineers to evaluate the standards and verify the methods.

When the standard time is used as a local performance measure, it allows the line employees to measure their performance. By comparing their performance to a standard for each task they perform, either they or management can evaluate whether changes need to be made in the job design.

Many organizations use the standard times to calculate efficiency reports. These reports are sometimes used by managers to identify areas to improve.

The efficiency report is a valid local performance measure, if higher machine efficiencies create higher profits for the firm. These reports are not valid local performance measures, if higher efficiencies of some work centers increase operating costs and/or inventory costs without increasing throughput. An example calculation is shown below

Efficiency is the ratio of actual units produced to the standard rate of production expected in a time period, or actual hours of production to standard hours, or actual dollar volume to a standard dollar volume in a time period. For example, if there is a standard of 100 pieces per hour and 780 units are produced in one eight-hour shift, the efficiency is 780/800 multiplied by 100, or 97.5%. Another term that some companies use is *operating efficiency*. This is defined by the *APICS Dictionary* as "a ratio (represented as a percentage) of the actual output of a piece of equipment, department, or plant as compared to the planned or standard output." For example, if the standard time for production of a part at your work center is 10 minutes and you have 420 minutes of capacity available at your work center (i.e., you can produce 42 parts in a shift), you would calculate your efficiency for an output of 50 parts during that 420 minutes as $(50/42)^*(100) = 119\%$.

Efficiency—A measure (as a percentage) of the actual output to the standard output expected. Efficiency measures how well something is performing relative to existing standards; in contrast, productivity measures output relative to a specific input, e.g., tons/labor hour. Efficiency is the ratio of (1) actual units produced to the standard rate of production expected in a time period, or (2) standard hours produced to actual hours worked (taking longer means less efficiency), or (3) actual dollar volume of output to a standard dollar volume in a time period.

APICS Dictionary, 9th edition, 1998

Operating efficiency—A ratio (represented as a percentage) of the actual output of a piece of equipment, department, or plant as compared to the planned or standard output.

APICS Dictionary, 9th edition, 1998

Caution! Local performance measures must be used with care. Remember that all of us change our behaviors in response to rewards and punishments. We recognize quickly when local performance measures are used to evaluate our performance and either rewards or punishments are given to us based on our performance. If managers use the incorrect performance measure, then they may have people doing work that does need to be done and avoiding work that needs to be done so that they can look good based on the local performance measures.

A second use of standard times is to estimate the amount of labor that is needed to produce the plant's products. For example, a planner may know that welding a worm gear for a grain elevator takes 90 minutes and that each welder who works 40 hours a week has 36 hours of capacity (allowing for breaks and lunches). The planner can calculate that one welder can produce 24 worm gears a week. So if the production of 240 worm gears a week is necessary, then 10 welders are required.

Supply Chain Performance Measures

The key question in supply chain management is how to coordinate the efforts of every firm in the supply chain and every employee of those firms. The coordination must provide ever-increasing amounts of value added to those customers willing to pay for it.

Performance measures drive behavior in any system. The selection of performance measures is crucial inside a firm and throughout the supply chain. Managers coordinate behavior of their employees and of their partners in the supply chain by the use of performance measures. It is through the use of measures that we are able to determine if we are making progress towards our goals.

Managers measure to improve productivity. Measurement is one step in the cycle of measure, evaluate, plan, improve, and then start measuring again.

The ideal performance measure pushes every firm in the supply chain and all employees in each firm to direct all of their efforts to increasing the amount of money made by everyone in the supply chain. The ultimate measure for each firm in the supply chain is their return on investment (ROI) or capital productivity. But, over what horizon should this be evaluated? For one quarter at a time or over a decade? The problem is that there is no perfect performance measure which will always push firms and their employees in both the short and long term to make the best decision for the long-term benefit of the supply chain. The choice of performance measures in both the firm and the supply chain must be monitored.

Given that all performance measures have potential problems, it helps to understand how all the measures selected are related to the firm's and the supply chain's financial success. Gable (1997) illustrated this in Table 3.3.

The relationships outlined in Table 3.3 help managers in all firms in the supply chain to understand how their decisions influence the supply chain

Table 3.3 Relationship of Supply Chain Variables to Financial Statements

Income Statement	Supply Chain Variables	Key Performance Measures
Net sales	Customer service	Revenue
Cost of goods sold	Purchasing/MRP	
	Production scheduling and control	
Selling and administrative expenses	Order processing	Logistics costs
	Transportation	
	Warehousing	
	Inventory control	
	Packaging	
	Support activities	
Interest expense	Inventory carrying cost	
Income before income tax		Logistics profit contribution

Balance Sheet	Supply Chain Variables	Key Performance Measures
Assets		
Cash and receivables	Order cycle time	Order shipment performance
	Order completion rate	Receivable/credits caused by system malfunction
	Invoice accuracy	
Inventories	Inventory	Return on Inventory Investment
Plant and equipment	Facilities and equipment	Return on assets
Liabilities	Purchasing/MRP	Minimization of premature commitments
Debt and equity	Financing options	Control of investment in inventory

and everyone's income statement. The key variables of supply chain performance measures related to the income statement are sales revenue, logistics costs, and the logistics profit contribution. For example, to improve sales revenue the supply chain members need to improve their customer service, while reducing their costs of goods sold. They would do this by improving their purchasing, material requirements planning, production scheduling, and control of operations.

As the supply chain matures, it should be able to improve these measures beyond what a firm could achieve on its own. As discussed earlier in this chapter, if a firm is forecasting in isolation from other firms in the supply chain, all the firms in the chain are making their own forecasts and consequently introducing large amounts of error into their planning. If the forecast of the retailer is shared with the other members of the supply chain, then all members can plan on the same basis and the forecast error can be limited to this one stage of the supply chain. As suggested by Table 3.3, this should allow the costs of goods sold to be reduced because purchasing and production scheduling will be more precise.

Role of Supply Chain Management Software

The rationale for investing in supply chain management software systems is the same as that required for investment in material requirements planning and enterprise resource planning software systems. The goal is to reduce inventory levels while improving customer service.

Improving your inventory flow internally will not benefit your firm, if it is not improved throughout the rest of the supply chain. The market for supply chain management software is growing rapidly. Many companies which offered only enterprise resource planning systems a few years ago, now offer web-based supply chain management software. This software allows suppliers and customers to communicate easily with each other. Information can easily be shared between computers, so that every member in the supply chain is communicating with the other members.

An Example of Supply Chain Management

Russell Stover Candies, the largest manufacturer of fancy boxed chocolates in North America, acquired a new enterprise resource planning (ERP) software package which included a supply chain management (SCM) solution. Russell Stover Candies operates six manufacturing plants and two box manufacturing plants in the United States. It also has 10 distribution centers in North America, Australia, New Zealand, and China. The supply chain is complex. It has over 2,000 suppliers and over 62,000 customers. Customers include mass merchandisers such as K-Mart and Wal-Mart and some small, independent drug stores. So, order entry and order processing is complicated and forecasting demand and scheduling is even harder.

The customer base could be segmented into two. One segment consists of the small customers, and the other is the large warehouse accounts which handle their own storage, distribution, and transportation. Each segment has its own unique requirements.

The small customers require a focus on logistics, ensuring that the product is delivered in a timely manner (chocolate is perishable with a short shelf life). The view of the supply chain with customers who handle their own storage (e.g., Wal-Mart) had to be extended to include the customers of the warehouse. At Wal-Mart this would be the individual stores. This requires working with the warehouse customer to improve the accuracy of the forecasts of the end customer.

Internal improvement of supply chain management focuses on maximizing production and resource utilization within Russell Stover to obtain a high return on assets. Russell Stover purchased and installed a new enterprise resource planning (ERP) system to help manage their own resources. The purpose of installing the new ERP system was to be able to use the information system in real-time. The new ERP system allowed Russell Stover to immediately record receipts and share this information across the firm. Their goal is to reduce their inventory within the plants, while providing better service to their customers.

4 Variance and Dependent Events

What Is Variance?

In industry the term *variance* can be used two different ways. First, there is variance between planned performance and actual results. Second, there is a statistical measure of central tendency of a set of data, which is called *variance*. It is this second or statistical definition of variance that is discussed here. For example, if the mean flow time in a factory is 10 days, but some jobs are completed as early as 3 days and others take as long as 25 days, there is variance in the flow-time data. Equation 4.1 is used to calculate the variance of a set of data. First, the mean is calculated (shown here as $\overline{X}$). Then the difference between the mean and each piece of data in the set is calculated $(X_i - \overline{X})$ and then this difference is squared. The sum of these n pieces of data squared is then divided by $(n - 1)$, which is one less than the number of data points.

$$\text{Variance } \sigma^2 = \frac{\Sigma (X_i - \overline{X})^2}{n - 1} \tag{4.1}$$

The square root of the variance is the standard deviation (σ), which is the measure most of us use when we are talking about the variance in a set of numbers. For example, it is useful to talk about the number of standard deviations a piece of data is from the mean of the distribution. The larger the number of standard deviations that a piece of data is from the mean, the less likely it is that that particular piece of data should really be considered part of the same population as the rest of the data. Figure 4.1 shows a particular distribution of a set of data. The distribution shown here is called the *normal distribution*. It is

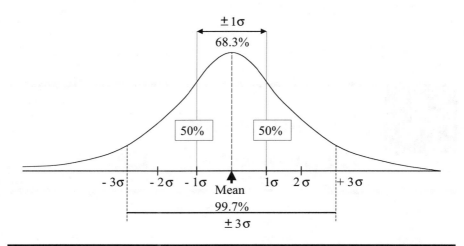

Figure 4.1   Normal Curve

sometimes referred to as a bell-shaped curve, since it resembles a bell. The normal distribution is very well understood and it is easy to analyze. An extremely important feature of the normal distribution is what is known as the *central limit theorem*. This theorem states that the averages of a set of samples from a population will be distributed normally. This means that if I take many small samples, the average of these samples has a normal distribution. A picture of the distribution of these averages will be bell shaped as in Figure 4.1.

Variance—1) The difference between the expected (budgeted or planned) value and the actual. 2) In statistics, a measure of dispersion of data.

APICS Dictionary, 9th edition, 1998

In the normal distribution, 50% of the population is above the mean or average and 50% of the population is below the mean. At any point, measured in the number of standard deviations above or below the mean, it is easy to see how much of the population is included. For example, the area under the curve that is one standard deviation above and below the mean is 68.3% of the population. Notice that 32.5% of the population is not within plus and minus 1 standard deviations of the mean. Half (16.25%) of this population is in the lower tail and half is in the upper tail of the bell-shaped curve (see Figure 4.1).

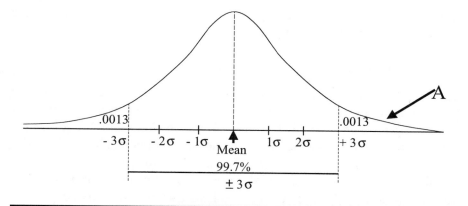

Figure 4.2   Normal Curve

At two standard deviations above and below the mean, 95% of the population is included. At three standard deviations, 99% of the population is included.

Statisticians evaluate the probability that a given piece of data is part of a certain population by measuring how many standard deviations it is from the mean. If a data point is more than 3 standard deviations from the mean, then we know that there is only a 0.26% chance that it came from the population being evaluated (i.e., the population represented by the normal curve). This is illustrated in Figure 4.2. Arrow A is pointing to the upper tail of the normal distribution. This tail is the area that is greater than 3 standard deviations from the mean. Since some points not within ±3 σ of the mean could be in the lower tail, there is only a 0.13% chance that the data represented by point A is from the same population as the other points which are under the curve in the area shown as ±3 σ from the mean.

Another measure of the variance of a data set calculates the relative size of the variance. It is called the coefficient of variation (CV). This is expressed mathematically in Equation 4.2 as:

$$CV = \frac{\text{Standard Deviation}}{\text{Mean}} \qquad (4.2)$$

The coefficient of variation gives the relative size of the standard deviation compared to the mean. So, when the standard deviation is 500 and the mean is 1, the CV is 500. But, when the standard deviation is 500 and the mean is 500, the CV is 1. This second example indicates that there is relatively little variance when compared with the first example. Another example is a measure of the mean time between failures (MTBF) of a machine. If the MTBF has a standard deviation of 100 and a MTBF of 500, then its coefficient of variation is 0.2. If the standard deviation of the MTBF is 100 and the mean is 100 hours, then its coefficient of variation is 1.

Dependent Events

All real-life systems have dependent events. When an event is dependent on another event it means that it cannot occur until the first event is completed. For example, on an assembly line, it is not possible to assemble two components onto a third part—the casing—unless the components and the casing have all been fabricated earlier. So, the assembly of the parts is dependent on the fabrication of the parts. Many of the manufacturing philosophies mentioned earlier (e.g., JIT and TOC) recognize the great influence of dependent events on the performance of the process. These manufacturing philosophies recognize that dependent events quickly spread disruptions throughout the process. So, both philosophies include methods of preventing delays from occurring in a sequence of dependent events. Techniques to do this will be discussed in later chapters.

A common illustration of serial or linear dependence is shown in Figure 4.3. In this illustration no events occur in parallel. For example, A must be completed before B and B completed before C. It is easy to see from the illustration that a delay at one event will delay all subsequent events. In this way delays will be spread throughout the system, as work center after work center falls behind waiting for the arrival of a job.

Interaction of Variance and Dependent Events

No system can be absolutely precise. There is variation in all human systems. All systems also have sets of dependent events. The combination of dependent events and variance can spread disturbances throughout the system if it is not controlled. For example, in the morning one work center may be starved for work because the work center feeding it is operating at a slower than average rate. But, in the afternoon the work center that was starved could become loaded with excess work when the work center feeding it finally speeds up and

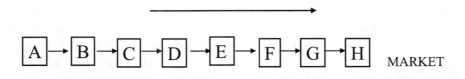

Figure 4.3   Dependent Events

completes its assigned tasks. Even if this second work center is operating well, the variance at the prior work center has created a situation where the second center has lost productive time, which cannot be recovered. So, the second work center is suddenly behind in its schedule. If this second work center has a large amount of spare capacity it is possible that it can quickly catch up, but if it does not have enough spare capacity, it will pass the delay caused by its inability to start the job on time onto the next machines in the job's route through the shop. Variance in one process is spread throughout the shop by the various series of dependent events that exist in all shops.

Placing spare capacity at strategic locations in a set of dependent events buffers the later stages of the process from variance that occurs earlier in the process. This spare capacity is referred to as *protective capacity*. It is not discussed in detail in this chapter but will be discussed in a later chapter. Techniques that can be used to reduce variance will also be discussed later.

Goldratt (1992) has repeatedly taught that understanding the interaction of dependent events and variance is essential to effectively manage any company. It is through the interaction of variance and dependent events that serious problems grow to the point where they can reduce the effective capacity of a firm. As a firm's capacity is wasted because of the interaction of variance and dependent events, its profits and cash flow are reduced.

Importance of Variance in the Supply Chain

A supply chain is a set of dependent events. Each firm in the chain depends on the firms that precede it and deliver its required inputs. If one of the firms does not deliver the product required by another firm on time, the second firm will be late producing and delivering its product to its customers. If the final customer of the chain has the option of getting the product elsewhere, he may decide to do so. So, a delay by one firm in a supply chain can result in a loss of business for all of the members of the supply chain.

This is illustrated in Figure 4.4, where Firm B has no variance in its production. It always takes exactly the same amount of time to make the product. But, its supplier, Firm A, has a lot of variance in its production. This variance is illustrated by the normal curve. When Firm A takes less time than expected to complete its work (i.e., the processing time is from the left side of the normal curve), the work arrives at Firm B early. But because Firm B has other work to do it probably cannot start its portion of the job early, so the input from Firm A sits until Firm B was scheduled to start that work. Since Firm B always takes the same amount of time to do its production, there will be no

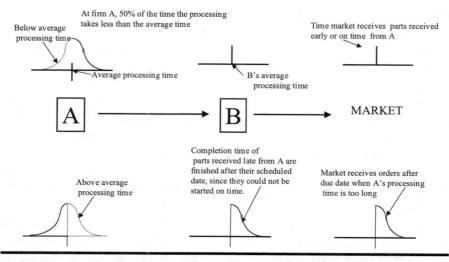

Figure 4.4   Accumulation of Statistical Fluctuation in a Supply Chain

delay in the delivery of these products to the final customer. They will be delivered right on time—neither early nor late—so the delivery distribution for the parts Firm A sent early to Firm B is a straight line.

However, when Firm A takes longer than average to get its work to Firm B (i.e., the processing time at Firm A is somewhere to the right side of the mean of the normal curve), the work arrives late to Firm B. Even though Firm B will take exactly the same amount of time to do its work, the shipment of the parts to the customer will be delayed. The final customer will perceive the products from Firm B as late. The delivery time distribution that the market uses to judge Firm B's performance shows that the product is never received early, it is received on time about 50% of the time but more importantly it is late 50% of the time. The final customer will of course blame Firm B, its supplier, even though the only mistake that Firm B made was to select Firm A as its supplier.

To compensate for the variance in its supplier's process, Firm B may take actions which will cost it money. Firm B may choose to hold inventory of raw materials from Firm A to buffer its processes from the variance in Firm A's process. Or, Firm B may choose to begin its production early, so that if it encounters a delay due to its supplier, this delay will not be passed onto its customer.

Any of these actions by Firm B to compensate for the variance in the processes of Firm A will cost Firm B money. There may be inventory costs for holding additional raw materials, or there may be inventory costs for holding the finished goods inventory until it is time to ship them to the customer. Additionally, Firm B may not be able to respond as quickly to its final customer because of its longer lead time.

5 Basics of Quality Management

The practices of quality management have evolved rapidly in the last 30 years. They will continue to evolve for the foreseeable future. One of the most basic practices of quality management is to ensure that all decisions are based on facts. This chapter will review the basics of quality management and identify the best practices. First, it is important to agree on what we mean when we say *quality*.

Definition of Quality

There are many definitions of *quality* in use. Juran, a famous teacher in the field of quality, states that there are two dominant meanings to the word *quality*. They are:

1. Quality consists of those product features which meet the needs of customers and thereby provide product satisfaction.
2. Quality consists of freedom from deficiencies. (Juran and Gryna, 1988, p. 2.2)

In this chapter *quality* means conformance to requirements, which agrees with the APICS definition shown below. The product features include goods and services. These are the properties possessed by the product bundle that are intended to satisfy the customers' needs. It is the customers who determine how satisfying the product is to them, based on their evaluation of the product compared with others in the market. A product that is free from deficiencies could be described as conforming to requirements or specifications. It is important that the specifications recognize the customer needs for the product so that meeting the specifications satisfies the customer. It is much easier to communicate within the firm in terms of specifications than it is in terms of the customer needs, so most firms use detailed specifications for the production or products and services.

Quality—Conformance to requirements or fitness for use. Quality can be defined through five principal approaches: (1) Transcendent quality is an ideal, a condition of excellence. (2) Product-based quality is based on a product attribute. (3) User-based quality is fitness for use. (4) Manufacturing-based quality is conformance to requirements. (5) Value-based quality is the degree of excellence at an acceptable price. Also, quality has two major components: (1) quality of conformance—quality is defined by the absence of defects, and (2) quality of design—quality is measured by the degree of customer satisfaction with a product's characteristics and features.

APICS Dictionary, 9th edition, 1998

Juran's Quality Trilogy

Juran stated that to create a firm that consistently provided good satisfaction to its customers, the manager had to constantly work through the quality trilogy, which is illustrated in Figure 5.1. The first step in this trilogy is quality planning. This occurs before the production of parts. Developing a quality plan is part of the product planning process. As the product planning process determines which market segment is to be served, it also determines which features the product must have to satisfy the customers and from that the quality specifications needed to satisfy the customer requirements. Given these customer requirements, a quality plan is developed that determines how the company will ensure that the product meets all of the customer requirements.

The second phase of the quality trilogy is quality control. In this step actual results are compared to the planned results. This is illustrated by the graph in Figure 5.1. The graph helps to answer the question of whether the products being produced really meet the specifications that were developed during the planning stage.

The third phase of the quality trilogy is quality improvement. It is not enough to meet the customer specifications. It is important in today's economy that a company continually improves. This is illustrated in Figure 5.1 by the new zone of quality control once an improvement is made. The quality

Quality Trilogy: Quality Planning, Quality Control, Quality Improvement

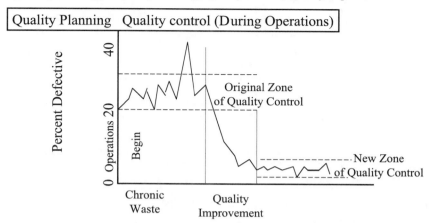

Figure 5.1 **The Juran Trilogy (From Juran and Gryna,** *Juran's Quality Control Handbook,* **McGraw-Hill, 1988, 2.7. With permission of the McGraw-Hill Companies.)**

trilogy continues after the improvement, to develop plans to maintain it and to control the process so that the improvement is not lost.

Total Quality Management

Total quality management (TQM) has many different definitions. But a definition that recognizes that TQM is still evolving was provided by Shiba, Graham, and Walden (1993). To them TQM is

> . . . an evolving system, *developed through success in industry,* for continuously improving processes, products and services to increase customer satisfaction in a rapidly changing world. (p.27)

Among the many different definitions of TQM, there are several common themes. These themes are that TQM focuses on the customer, involves everyone in the organization, and strives for continuous improvement. The TQM theme of customer focus recognizes that the purpose of the firm is to make a profit by satisfying the customer's needs. The TQM theme of involving everyone in the organization in the activities of TQM recognizes that customer satisfaction requires that everyone in the organization must become committed to satisfying the customer. The third TQM theme of striving for continuous improvement recognizes that customer satisfaction requires that the market is constantly changing so that the firm's product is never good enough. It can always be improved, and everyone is responsible for achieving this improvement.

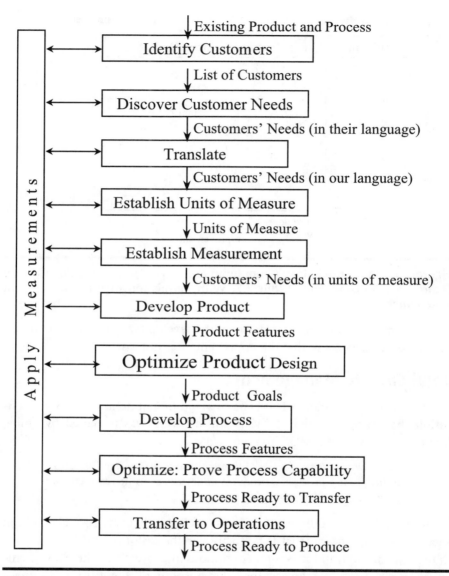

Figure 5.2 Quality Plan (From Juran and Gryna, *Juran's Quality Control Handbook,* McGraw-Hill, 1988, 6.5. With permission of the McGraw-Hill Companies.)

Total quality management—A term coined to describe Japanese-style management approaches to quality improvement. Since then, total quality management (TQM) has taken on many meanings. Simply put, TQM is a management approach to long-term success through customer satisfaction. TQM is based on the participation of all members of an organization in improving processes, products, services, and the culture they work in. The methods for implementing this approach are found in teachings of such quality leaders as Philip B. Crosby, W. Edwards Deming, Armand V. Feigenbaum, Kaoru Ishikawa, J. M. Juran and Genichi Taguchi.

APICS Dictionary, 9th edition, 1998

The approach to TQM can be systematic according to Juran (1988). The "Quality Planning Roadmap" in Figure 5.2 is centered around the need for measurements at each step to assure total alignment with customer demands. Each output becomes the input for the next step. By following such a systematic approach, each element of the supply chain is assured to be in alignment with the customers' quality requirements.

To successfully implement TQM, it must not become a program which requires employees to do additional work after their job is completed. Rather, TQM must become the way that work is done. The method of management is TQM.

Evolution of Quality Management

Since 1924, when Shewhart first released his work adapting the scientific method to quality control, the development of quality management practices has been driven by increased application of the scientific method. The scientific method is shown in Figure 5.3 as the Theory–Hypotheses–Data–Verification cycle. Simply stated, the scientific method says that theories must be stated in a way that hypotheses can be formulated which can prove the theory wrong. A hypothesis is proved wrong by obtaining data. This data is analyzed (usually using

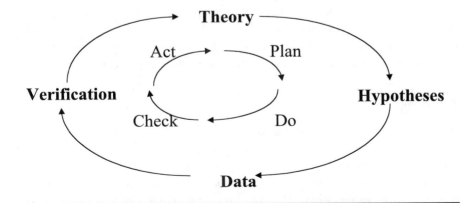

Figure 5.3   Scientific Method and PDCA

statistical tools) to verify whether the theory is correct or incorrect. For example, if we have a problem where 20% of the crystals that we are manufacturing are broken in the last stage of polishing, we could develop a theory that most of the defects are due to inattention by the workers who are in charge of the operation. We could then develop a hypothesis that our "better" operators would have fewer defects and another hypothesis would be that the defects would not depend on the material received at the work station by the operator. Given these hypotheses we could then collect data about the causes of the defects. To create a fair test, we may have different operators process the parts on different machines, and each operator may be given a variety of crystals from different batches. When we analyze this data if we found that all of the operators who performed the process on machine 2 had approximately the same number of defects and that these operators had a significantly lower number of defects when they performed the process on other machines, we would have grounds for rejecting our hypotheses that the workers are responsible for the defects. Instead we would have to revise our theory to recognize that poorly maintained machines create defects.

The difficulty with a direct application of the scientific method is that it is scary to most of us. Shewhart recognized this and created simple tools to apply the scientific method and statistically analyze the data. These basic quality management tools are used by line workers and quality practitioners throughout the world today. These basic tools allow everyone in the organization to manage by using facts.

One of these tools is the Deming Cycle. The Deming Cycle is also known as the Shewhart Cycle or PDCA. The initials PDCA stand for Plan-Do-Check-Act. This concept is illustrated in Figure 5.3 as the inner circle. The PDCA cycle is the scientific method phrased differently. During the Plan step, issues or problems are examined analytically and quantitatively and a plan of action is developed. The plan is then implemented during the Do step and data is gathered. During the Check step, data is analyzed quantitatively and qualitatively to determine whether the plan worked. And, at the Act step the previous procedures and plans are modified to incorporate the successes from this cycle. This may require standardizing a new procedure or changing equipment.

Deming

W. Edwards Deming is recognized as one of the gurus of total quality management. Deming focused on reducing variance in all processes of an organization. This included manufacturing processes as well as management, engineering, and all other processes. Deming believed that variation accounted for most problems. He believed that a primary requirement of good management is that the manager be able to determine whether the existing variance was due to common causes or to assignable causes (i.e., special causes).

Assignable cause—A source of variation in a process that can be isolated, especially when its significantly larger magnitude or different origin readily distinguishes it from random causes of variation. Syn: special cause.

APICS Dictionary, 8th edition, 1995

Common causes—Causes of variation that are inherent in a process over time. They affect every outcome of the process and everyone working in the process.

APICS Dictionary, 8th edition, 1995

The difference between special or assignable cause and common cause is illustrated in Figure 5.4 using a normal curve. Typically, managers assume that any point outside of $\pm 3\ \sigma$ of the mean is not part of the population represented by the normal curve. These points can then be investigated further and a specific cause can be assigned as a reason for why that point deviates so greatly from the mean. The points in either tail of the curve are all treated as assignable points. All the points that are within the $\pm 3\ \sigma$ of the mean are common cause.

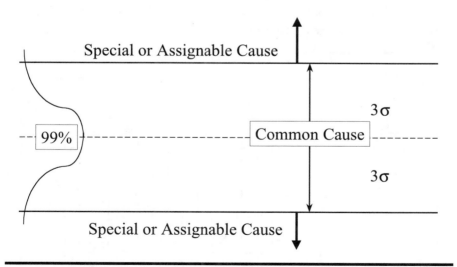

Figure 5.4   Special or Assignable Cause

Variance within the common cause area is due to the system and cannot be assigned to a particular cause. Only by changing the system can we decrease the amount of common cause variance.

A common set of tools used by managers and workers to identify and reduce variance are referred to as the 7 Quality Control Tools. They are explained and discussed in the next section.

The 7 Quality Control Tools

The 7 Quality Control Tools are basic quality tools that allow data to be collected and factually analyzed. They are listed in Table 5.1.

Table 5.1   The 7 Quality Control Tools

Checksheets
Pareto Diagrams
Cause-and-Effect Diagrams
Graphs
Control Charts
Histograms
Scatter Diagrams

Checksheets

Checksheets are used to gather data and compile it in such a way that it can be easily analyzed. They are often used to gather information about a production process or about defects. Whenever possible the checksheet is designed so that the data is sorted as it is collected. This is illustrated in Figure 5.5. This allows

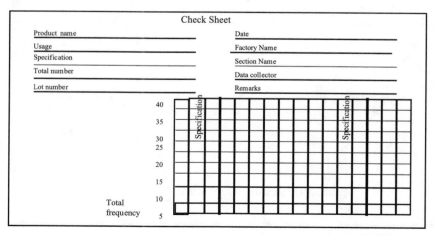

Figure 5.5   Checksheets

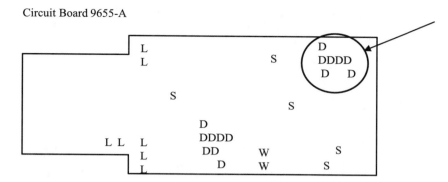

L = Lamination problem
W = wrong component
D = defective component
S = solder problem

Figure 5.6   Location Checksheet

whoever is gathering data to record the dimensions of the product being measured as a check mark in the appropriate column.

There are many types of checksheets in use. A different type of checksheet is shown in Figure 5.6. This shows a picture of a circuit board. During the shift whenever a defect is found, the type and location of the defect is recorded onto the sheet. For example, the Ds show where a device was not installed. The circle in Figure 5.6 encloses a group of Ds in the upper right corner. The placement of the defects provides a visible clue for the worker to begin investigating what defective component is being placed onto the board.

Pareto Diagrams

A Pareto diagram is a bar chart, which is usually used to suggest which problem should be solved first. What distinguishes a Pareto diagram from a typical bar chart is that each classification is ranked in descending order. So, the defect with the greatest number of occurrences is arranged on the left. A Pareto diagram is illustrated in Figure 5.7. In this diagram the defect categories are given on the bottom on the x axis. Notice that the defects are arranged from left to right in descending order. The left vertical axis or the y axis shows the number of defects. The right vertical axis shows the cumulative percentage. In this example, caulking was the cause of 198 defects, or 47% of the total defects. So, the right vertical axis shows that the cumulative percentage line for this defect to be 47%. The second largest category of defects was

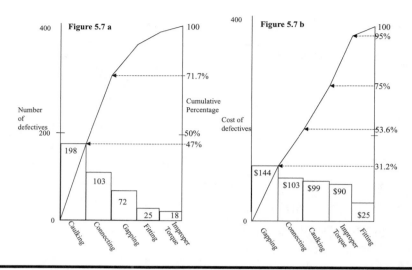

Figure 5.7   Pareto Diagram

connecting. It had 103 defects, which was 24.7% of the total number of defects. So, the cumulative percentage of defects for caulking and connecting together is 71.7% as shown by the graph on the right hand vertical axis of Figure 5.7a.

The Pareto diagram is a valuable tool for communication. It quickly demonstrates the importance of a problem being discussed. Sometimes the information is translated into dollars to improve communication. This can be important in many situations where the costs of the defects vary greatly. For example, how much does a caulking defect cost? How much does a connecting defect cost? In Table 5.2, the cost of each type of defect is shown. The costs may be based on the amount of time to fix the defect during final inspection, where they were found. The new Pareto diagram using this cost data is given in Figure 5.6. Now the largest defect is gapping, followed by connecting.

Table 5.2   Cost of Defects by Type

	Number of Defects	Cost of Defects	Extended Cost	Percentage
Caulking	198	$0.50	$ 99	21.48%
Connecting	103	1.00	103	22.34%
Gapping	72	2.00	144	31.24%
Fitting	25	1.00	25	5.40%
Improper torque	18	5.00	90	19.52%

To create a cause-and-effect diagram to sort out causes and identify relationships
1. Determine quality characteristic with variance (be sure to state as a weakness)
2. Write main factors on branches grouped by materials, equipment, methods, people, and other
3. Brainstorm all possible causes in each category and record detailed factors on each branch. Have at least 5 levels including the problem statement (i.e., 1) for each possible cause

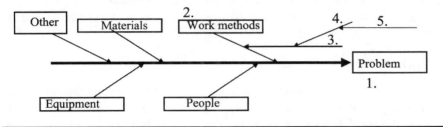

Figure 5.8   Cause-and-Effect Diagram

Cause-and-Effect Diagrams

The cause-and-effect diagram (CED) is also known as the Ishikawa diagram. It is used to sort out the causes of a defect. It is recommended that it be used with a team of individuals who are trying to solve a weakness or a problem. As shown in Figure 5.8 the problem is written in the box on the right of the page. A line (like the backbone of a fish) comes from the problem in the box to the left side of the page. Separate lines for each group of related factors come off of this backbone. In most work situations it is common to group factors into the categories of material, work methods, equipment, people, and other as shown in Figure 5.8. As causes of the problem are given, they are grouped into one of these categories. The facilitator then asks "why." Either the person contributing the possible cause of the problem or someone else in the group suggests why an item causes a problem. That reason is then written on an arrow attached to the item that generated the question. The facilitator pushes the group to take each possible cause to at least 5 levels of *why*. This is shown in Figure 5.8 for the category of work methods. The first *why* level is the problem statement. The second *why* level is the major work methods. The third *why* level is labeled with 3. And the fourth and fifth *why* levels are labeled with 4 and 5, respectively.

A CED is used to communicate the possible causes of a problem to everyone. One important part of this communication is to get ideas from as many people as possible. Everyone who is available is asked, "What are the causes of this problem?" The effort of building the CED teaches everyone involved new knowledge as well as solving the problem.

Once the CED is created, the team actively gathers data to investigate each of the causes. The team then analyzes this data to determine what the primary causes of the problem are. Remember that the CED was created to identify all possible causes and it typically includes many items that may not be contributing to the problem; so once it is created the actual cause must be separated out from those on the list. To do this, some of the other 7 Quality Control Tools can be used to help analyze the CED.

Graphs

Graphs can be an effective communication tool. To be effective they must organize and summarize the data simply. To do this you must know why you are creating the graph before you do it. One of the simplest graphs is a line graph. This is often used to represent the progress of a group or of a plant over time as shown in Figure 5.9. This shows the production each day at a glance.

A graph is a picture. Like other pictures, a graph receives the most attention when it is shared with everyone and when there is a change in it. As long as the line on the graph in Figure 5.9 moves within the same range up and

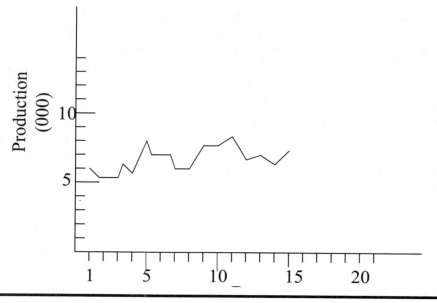

Figure 5.9   Line Graph

down, it will not be noticed. However, if production jumped above 10,000 units in one day, the level of change would be very noticeable and would help focus employee efforts on identifying what changed.

Control Charts

Control charts are a special type of graph. They not only graph some value over time, but they also analyze how the data moves. The purpose of this type of graph is to determine if the data is normal or abnormal. Or in Deming's terminology, they help to separate assignable causes from common causes. There are many types of control charts, but only one type will be explained here. The Xbar and Rbar charts are common control charts that are always used together and they will be explained here. These control charts are used with continuous data or data that is not whole numbers. For example, the width of a sheet of typing paper could be tracked using an Xbar chart. In the United States, typing paper is usually $8\frac{1}{2}$ inches wide or 8.50". In some cases, though, the paper may actually be only 8.499" wide, or it may be as wide as 8.501".

If our plant cuts rolls of paper into typing paper, we would be very concerned that we cut the paper very close to the stated size. To track how effectively we do that, we would gather data over time. Typically we would gather the same amount of data every time. For example, we may always take a sample of 5 sheets of paper and we may always take a sample after every cut of 100,000 sheets. Six samples are shown in Figure 5.10 at the top of the graph.

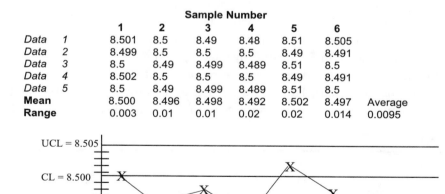

		Sample Number						
		1	**2**	**3**	**4**	**5**	**6**	
Data	*1*	8.501	8.5	8.49	8.48	8.51	8.505	
Data	*2*	8.499	8.5	8.5	8.5	8.49	8.491	
Data	*3*	8.5	8.49	8.499	8.489	8.51	8.5	
Data	*4*	8.502	8.5	8.5	8.5	8.49	8.491	
Data	*5*	8.5	8.49	8.499	8.489	8.51	8.5	
Mean		8.500	8.496	8.498	8.492	8.502	8.497	Average
Range		0.003	0.01	0.01	0.02	0.02	0.014	0.0095

Given that the center line (CL) is 8.500 and A2 for a sample of 5 is .577 then
UCL = 8.500 + A2(.0095) = 8.500 + .577(0.0095) = 8.505 and
LCL = 8.500 - A2(.0095) = 8.500 - .577(0.0095) = 8.495

Figure 5.10   Xbar Control Chart

There are 5 pieces of data in each sample, which is in the column below the sample number. At the bottom of each sample the mean of the sample (i.e., Xbar) is shown. And below that the range is shown.

Each of the sample means is placed onto the graph below the table as shown in Figure 5.10. The graph also has on it the upper control limit (UCL) and the lower control limit (LCL). These are calculated as shown in Figure 5.10. The average range is multiplied by A2 (see Table 5.3). This product is then added to the nominal center of the process. The nominal center of the process is what the process should actually be doing. In this case, we should actually be cutting all the paper to be 8.50″ in width. If we do not have a nominal value, then we take the grand average of all of the samples to find the center line (CL). The UCL, LCL, and CL are always clearly labeled on the control chart. This makes it easy for everyone to read and understand the chart. In this graph one of the points on the Xbar chart is below the LCL. This means that the process is not in control. There is something abnormal about the process. The operator is expected to stop the process and find out what is causing the error or call for help. In Deming's terminology, points outside of the control limits indicate an assignable cause, while those points inside the control limits represent variance that is due to common causes.

The Xbar and the Rbar graphs are always created together. They show two different interpretations of the data. The Xbar chart shows whether the average value of the process has shifted from where it should be (in this case that is 8.500″). The Rbar chart shows whether the amount of variance in the process is increasing over time. To create the Rbar chart, place each of the sample ranges into the table as shown in Figure 5.11. The graph also has on it the upper control limit (UCL) and the lower control limit (LCL). These are calculated as shown in Figure 5.11 by looking up the numbers D3 and D4 in Table 5.3. The D4 is multiplied by the average range to find the upper control limit for the Rbar chart. The nominal center of the process is what the range should actually be or the grand average of the range data we have. The UCL,

Table 5.3

n	A_2	D_4	D_3
2	1.880	3.267	0.000
3	1.023	2.575	0.000
4	0.729	2.282	0.000
5	0.577	2.115	0.000
6	0.483	2.004	0.000
7	0.419	1.924	0.076

Adapted from Melnyk & Denzler, 1996, p. 357.

		Sample Number						
		1	**2**	**3**	**4**	**5**	**6**	
Data	1	8.501	8.5	8.49	8.48	8.51	8.505	
Data	2	8.499	8.5	8.5	8.5	8.49	8.491	
Data	3	8.5	8.49	8.499	8.489	8.51	8.5	
Data	4	8.502	8.5	8.5	8.5	8.49	8.491	
Data	5	8.5	8.49	8.499	8.489	8.51	8.5	
Mean		8.500	8.496	8.498	8.492	8.502	8.497	Average
Range		0.003	0.01	0.01	0.02	0.02	0.014	0.0095

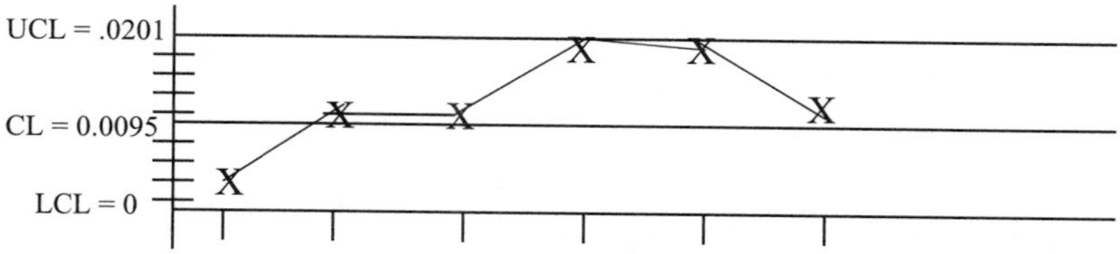

Given that the center line (CL) is 0.0095 and D4 and D3 for a sample of 5 are .2.115 & 0
UCLR = D4(.0095) = 2.115(0.0095) = 0.02009 or 0.0201 and
LCLR = D3(.0095) = 0(0.0095) = 0

Figure 5.11   Rbar Control Chart

LCL, and CL are always clearly labeled on the control chart. This makes it easy for everyone to read it and understand it. In this example, all of the sample points stay within the UCL and the LCL. But, there is something suspicious about the range anyway, since 5 of the 6 samples have ranges above the CL. This may indicate that the variance is increasing. Since the Xbar went out of control on the third sample, we say that the process was out of control.

Histograms

A histogram is a bar chart that is used to make a picture of the data's variance. It lets us examine if there is something wrong about a particular process. Histograms are often used to establish whether a process has a normal- or bell-shaped curve. A histogram shows the central location, the shape, and the spread of the data. It is very useful to help us visualize the characteristics of the distribution.

Questions that can be answered by the histogram are: What is the shape of the distribution? What is the mean of the distribution? How much dispersion of the data is there? Is the distribution symmetrical? Is the distribution skewed? Is there only one peak? Is the distribution cliff-like? Does the distribution look like a cogwheel? What is the relationship of the distribution with the customer's specifications?

Data used to prepare the histogram in Figure 5.12 are shown in Table 5.4. The first step in preparing a histogram is to find the range of the data (i.e., the maximum value – the minimum value). In Table 5.4, the maximum value is 3.68 and the minimum is 3.30, so the range is 3.68 − 3.30 = 0.38. Over years of study, statisticians have found that the best histograms are produced when the number of buckets or cells is within the range given in Table 5.5. In this example, there are 100 pieces of data. So, we can have 6 to 10 cells.

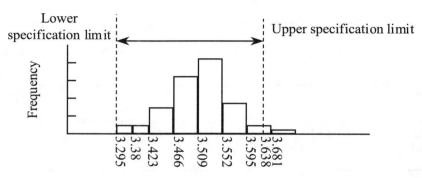

Figure 5.12   Histogram Example (Adapted from Ishikawa, 1982, p. 7)

Table 5.4   Sample Data for Histogram

3.48	3.56	3.50	3.47	3.52	3.48	3.46	3.50	3.56	3.38
3.41	3.37	3.47	3.49	3.45	3.44	3.50	3.49	3.46	3.46
3.55	3.52	3.44	3.50	3.45	3.44	3.48	3.46	3.52	3.46
3.48	3.48	3.32	3.40	3.52	3.34	3.46	3.42	3.30	3.46
3.59	3.63	3.59	3.46	3.38	3.52	3.45	3.48	3.31	3.46
3.40	3.54	3.46	3.51	3.48	3.50	3.68	3.60	3.46	3.52
3.48	3.50	3.56	3.50	3.52	3.46	3.48	3.46	3.52	3.56
3.52	3.48	3.46	3.45	3.46	3.54	3.54	3.48	3.49	3.41
3.41	3.45	3.34	3.44	3.47	3.47	3.48	3.54	3.47	3.41
3.40	3.54	3.46	3.51	3.48	3.50	3.68	3.60	3.46	3.52

Table 5.5   Selecting Classes for Histograms

Number of Data	Number of Cells
Under 50	5–7
50–100	6–10
100–250	7–12
Over 250	10–20

From Ishikawa, 1982, p. 8.

To illustrate, let us pick 8 cells for the histogram. The cell width will be (0.38/8) = 0.0425, or 0.043. Starting the first cell width one significant digit smaller than the minimum value (i.e., 3.295), the cell widths are shown in Figure 5.12.

Scatter Diagrams

Scatter diagrams are among the easiest graphs to make. They are used to examine the relationship between cause and effect. For example, if someone proposes that the variance in the length of rubber being cut on machine 2 is due to the drive belt loosening during production, we might take a series of measurements to establish whether this is indeed the cause. The measurements that were taken are shown in Table 5.6. The length of the rubber being cut is shown on the y axis of Figure 5.13. The deflection in the drive belt is shown on the x axis.

By examination of the scatter diagram we might determine that there is a linear relationship between the deflection in the drive belt and the length of

Table 5.6   Data for Scatter Diagram

Sample	Deflection (cm)	Length (mm)	Sample	Deflection (cm)	Length (mm)	Sample	Deflection (cm)	Length (mm)
1	1.5	1100	21	1.88	1108	41	1.83	1106
2	1.51	1101	22	1.92	1108	42	1.76	1107
3	1.49	1101	23	1.48	1100	43	1.87	1106
4	1.52	1100	24	1.53	1101	44	1.96	1107
5	1.79	1104	25	1.525	1101	45	1.91	1108
6	1.81	1104	26	1.55	1100	46	1.79	1106
7	1.78	1105	27	1.53	1100	47	1.83	1106
8	1.82	1104	28	1.51	1101	48	1.76	1107
9	1.83	1105	29	1.48	1101	49	1.87	1106
10	1.85	1106	30	1.52	1100	50	1.76	1107
11	1.86	1106	31	1.509	1100	51	1.69	1101
12	1.87	1107	32	1.51	1101	52	1.76	1105
13	1.845	1106	33	1.48	1101	53	1.88	1108
14	1.9	1107	34	1.52	1100	54	1.96	1109
15	1.91	1108	35	1.61	1105	55	1.67	1105
16	1.95	1109	36	1.69	1104	56	1.79	1106
17	1.92	1108	37	1.65	1105	57	1.68	1106
18	1.93	1108	38	1.63	1104	58	1.76	1107
19	1.91	1108	39	1.68	1105	59	1.71	1100
20	1.89	1109	40	1.79	1106	60	1.81	1107

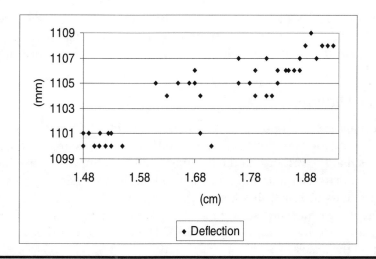

Figure 5.13   Scatter Diagram

the rubber being cut. This graph suggests that if we want a more uniform length of rubber, we need to limit the deflection of the drive belt.

If there is no evident correlation, Ishikawa suggests that the data being graphed be stratified. For example, if this is data from two shifts then the data for the first shift would be graphed separately from the data for the second shift.

Correlation—The relationship between two sets of data such that when one changes, the other is likely to make a corresponding change. If the changes are in the same direction, there is positive correlation. When changes tend to occur in opposite directions, there is negative correlation. When there is a little correspondence or random changes, there is no correlation.

APICS Dictionary, 8th edition, 1995

If there is the possibility of some correlation (i.e., a relationship between the two sets of data), Ishikawa recommends using a simple sign test to see if the correlation is significant. This test is illustrated in Figure 5.14.

1. Calculate the median values for both the *x* and *y* axes. Put these lines onto the graph. This divides the scatter diagram into four quadrants. In this example the median for the x axis was 1.77 cm deflection and the median for the y axis was 1106 mm length.

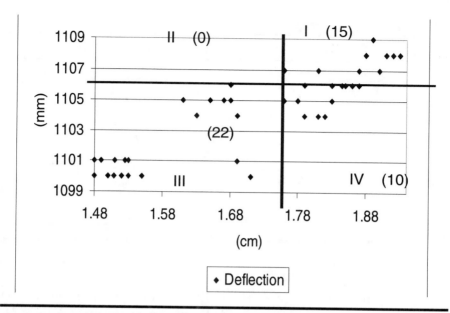

Figure 5.14   Correlation Test

Median—The middle value in a set of measured values when the items are arranged in order of magnitude. If there is no single middle value, the median is the mean of the two middle values.

APICS Dictionary, 8th edition, 1995

---

2. Number each quadrant counterclockwise beginning with the quadrant in the upper right corner (i.e., between 12:00 and 3:00 on the clock). Count the number of points in each quadrant and the points on the line. These are recorded on the example in Figure 5.14. (Note: double points or points on top of another point are counted.)

3. Total the number of points in quadrants II and IV and the grand total of points for all 4 quadrants less the number of points on the line. In this example, the total of points in quadrants II and IV is $0 + 10 = 10$. The number of points in quadrants I and III is $15 + 22 = 37$. The number of points on the line is 13, so the total number of data points minus the number of the points on the line is $60 - 13 = 47$.

4. Compare the total number of points in quadrants II and IV (i.e., 10) to the limit on the number of points given in Table 5.7. Since $N = 47$, the limit is 16, so II + IV < 16. Compare the total number of points in

Table 5.7   Sign Test Table

N	Limit on Number of Points		N	Limit on Number of Points	
	II + IV at 5%	I + III at 5%		II + IV at 5%	I + III at 5%
30	9	21	40	13	27
31	9	22	41	13	28
32	9	23	42	14	28
33	10	23	43	14	29
34	10	24	44	15	29
35	11	24	45	15	30
36	11	25	46	15	31
37	12	25	47	16	31
38	12	26	48	16	32
39	12	27	49	17	32

From Ishikawa, 1982, p. 93.

quadrants I and III (i.e., 37) to the limit on the number of points in the table. In this example, for N = 47 the limit is 31, because I + III > 31. If there was no correlation the total of I + III and II + IV would be almost equal. In this example II + IV are significantly less than expected by chance and I + III are significantly greater than expected by chance, so there is correlation present. Also, since (I + III) > (II + IV) the correlation is positive.

Summary

These 7 Quality Control tools are commonly used management tools. They are simple, but they take data, produce information quickly, and help present the information in a way that others can take effective action. They are used not only for quality management, but also for management of other tasks such as supply chain management.

6 | Operating Environments

T his chapter provides a basic understanding of the different characteristics that firms develop to compete in different market environments.
These operating environments form constraints on all of the operations decisions that are made and must be understood by everyone working
in the environment. These characteristics influence the firm's supply chain
management decisions.

Customer Order Lead Time

All companies operate in different environments because they provide different customer needs. The length of time the customer is willing to wait for delivery of the product from the time they order the product to receiving the
product is one of these characteristics. The differences in terms of lead time
for these operating environments are shown in Figure 6.1.

When Motorola orders a satellite from Ford Aerospace, it expects to wait
until the satellite is designed, built, and tested before taking delivery. This type
of purchase is an Engineer-to-Order (ETO) purchase. The ETO firm sells its
customers its capabilities in certain areas. A common example of this is the

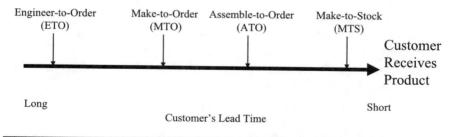

Figure 6.1   Environments and the Customer's Lead Time

civil engineering firm that helps design and construct a parking garage. In Figure 6.1 you can see that purchasing from an ETO firm requires the customer to accept a longer lead time than when purchasing from other types of firms.

The Make-to-Order (MTO) firm has designed a set of standard products, but the firm has not made the components in advance. Once a customer order is received for one of its standard products, the MTO firm begins to manufacture the product and ships the order when it has completed it. An example of a firm in this type of business might be one that produces cables and connectors for the computer industry. This type of firm would have a large catalog of standard cables and connectors. Its customers would call and place an order for a particular type of cable or connector. Once it had this order, the company would then manufacture and ship the product to the customer when completed. This firm has a much shorter lead time than the ETO firm. An advantage to the firm is that it does not carry any partially assembled product and does not have to forecast customer demand for specific end products. It manufactures only those products for which it has firm orders.

The Assemble-to-Order (ATO) firm has designed a product which is manufactured from a variety of components. These components can be assembled into a wide range of products. The company stocks the components, but does not assemble the product until it receives a customer order. The customer is allowed to select from a variety of standard options, which are then assembled into the final product at the factory. A current example of this type of product is a computer manufacturer who waits until an order is received from a customer over the phone and then assembles the computer to meet the customer's specifications. This type of product has a fairly short lead time, but it still allows the customer to customize the product to suit his/her own unique needs. The manufacturing firm limits its risk to the amount of inventory that is stored as components for assembly.

The Make-to-Stock (MTS) firm designs and either assembles or manufactures a product that is shipped either to a finished goods warehouse or to a retail store before the customer wants it. This type of product is a standard product and often one that a customer considers either a convenience item or a commodity. There are many examples of these types of products. One example is hand tools. The typical customers for hand tools expect to walk into a retail store, select the tools they want, pay for them, and carry them home. The lead time for the customers is short. It involves only their trip to the store and back. Because the customers cannot have the tools customized to suit their needs, the firm's product designers must work very hard to identify those features and characteristics that the customers want before manufacturing the tools. The manufacturing firm must also forecast customer demand to have

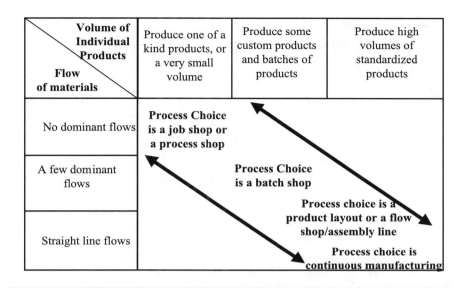

Volume of Individual Products / Flow of materials	Produce one of a kind products, or a very small volume	Produce some custom products and batches of products	Produce high volumes of standardized products
No dominant flows	**Process Choice is a job shop or a process shop**		
A few dominant flows		**Process Choice is a batch shop**	
Straight line flows		**Process choice is a product layout or a flow shop/assembly line** **Process choice is continuous manufacturing**	

Figure 6.2   Volume and Process Choice (Adapted from Hayes and Wheelwright, 1979, p. 133–140)

adequate supplies on hand before the customers require them, because it is likely that the customers will go to a competitor if a firm is out of stock on a particular item.

Process Choice and Layout

Firms are commonly classified by the methods that they select to manufacture their products. Usually the volume of product that it is producing determines the methods used by a firm to manufacture the product. As discussed earlier, Hayes and Wheelwright (1984) illustrated this relationship, which is reproduced in Figure 6.2. The common classifications are job shop, batch processing, repetitive manufacturing or line manufacturing, and continuous manufacturing.

In Figure 6.2, once you know the volume of production (given across the top), you drop a line straight down to the diagonal. This suggests the type of process that can be competitively used to produce the product. When producing a single product, or a very small number of products, a line from the top down indicates that the best process method is a job shop or process shop. This type of shop is organized so that all similar equipment is grouped together. The best type of physical layout of the equipment given this process choice is found by drawing a line from the process method to the left.

Flow Shop Layout

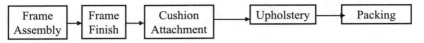

Process-Oriented or Functional Layout

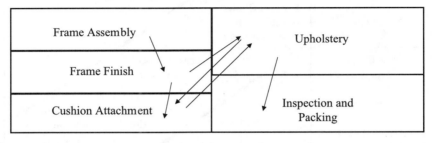

Figure 6.3   Factory Layouts for Chair Manufacturing

When the process choice is a job shop, the layout choice is also a job shop or functional layout. This type of layout makes economic sense when every order through the shop has a different routing (i.e., the machining sequence for each job is different). As production volume for each order increases, a line from the top in Figure 6.2 down indicates that batch processing is the most economical processing method. Batch processing is nothing more than grouping parts together and processing them at one time as a single lot. The advantage of batch processing is that the time that needs to be spent on setting up individual orders can be reduced as many small orders are combined into one large order. Drawing a line from the process method to the left shows that both a batch process and a job shop have a process or functional layout. The concept of a functional layout is illustrated in the lower part of Figure 6.3. When this chair factory uses a functional or process layout, the part moves from machine to machine as it needs different types of processing and it may actually return to machines visited earlier to receive more processing. Often in this type of shop, when the material flow through the factory is traced it begins to look like spaghetti.

Functional layout—A facility configuration in which operations of a similar nature or function are grouped together; an organizational structure based on departmental specialty (e.g., saw, lathe, mill, heat treat, and press).

APICS Dictionary, 8th edition, 1995

Job shop—An organization in which similar equipment is organized by function. Each job follows a distinct routing through the shop.
APICS Dictionary, 8th edition, 1995

Batch processing—A manufacturing technique in which parts are accumulated and processed together in a lot.
APICS Dictionary, 8th edition, 1995

Flow shop—A form of manufacturing organization in which machines and operators handle a standard usually uninterrupted, material flow. The operators generally perform the same operations for each production run. A flow shop is often referred to as a mass production shop or is said to have a continuous manufacturing layout.
APICS Dictionary, 8th edition, 1995

As volume increases, a line drawn from the top of Figure 6.2 to the process method suggests that the best process choice is product or flow shop or assembly line. Drawing an arrow to the left suggests that a product or flow layout is appropriate. Sometimes shops using this type of layout are called flow shops. This type of layout becomes possible when the shop has sufficient volume of the same product to be able to dedicate equipment to producing just that one product. Since the operators perform the same operations for each production run, it is cheaper to train the operators and they can become highly proficient at their assigned tasks, so the unit cost of production is reduced. This type of layout is illustrated at the top of Figure 6.3, where the material to produce a chair flows in a straight line as it flows through the shop.

The largest volume of production is at the extreme right of Figure 6.2. Drawing a line from here to the process method suggests that continuous manufacturing is appropriate. Continuous manufacturing usually refers to products that are not discrete. For example, gasoline is usually produced in a continuous process. The refinery does not produce one gallon at a time. Instead, it produces thousands of barrels in a continuous stream, so it is not possible to know where one gallon of gasoline begins and another ends until the gasoline is packaged. Because there is no discrete beginning or ending, this is sometimes referred to as lotless production. An important characteristic of continuous production is that the routing of the jobs is fixed, and setups are seldom changed. When individuals speak of continuous production in the context of mass production they are usually referring to shops with automated assembly lines. This term indicates that the material flow is continuous during the production process, that the routing of the jobs is fixed, and that setups are seldom changed.

Continuous production—A production system in which the productive equipment is organized and sequenced according to the steps involved to produce the product. This term denotes that material flow is continuous during the production process. The routing of the jobs is fixed, and setups are seldom changed.

APICS Dictionary, 8th edition, 1995

Mass production—High-quantity production characterized by specialization of equipment and labor.

APICS Dictionary, 8th edition, 1995

Continuous manufacturing usually refers to products that are not discrete. The *APICS Dictionary* defines continuous flow (production) as "Lotless production in which products flow continuously rather than being divided." However, sometimes individuals speak of continuous production in the context of mass production. When they speak of continuous production in this way, they are usually referring to shops with automated assembly lines. A common example of the assembly line is the automobile assembly plant.

At this point in the design of the production process we have an estimate of the volume (top of Figure 6.2), and we know which layout is appropriate. Next, we need to determine the capacity that we need to have available in the process.

Capacity Decisions

Capacity is one of the basic concepts of business. It is vital that a company understands its capacity characteristics to be able to compete. The first characteristic of capacity that must be understood is how much capacity is available. The *APICS Dictionary* defines *capacity available* as the "capability of a system or resource to produce a quantity of output in a particular time period." This is determined by the design of the shop.

Capacity—1) The capability of a system to perform its expected function. 2) The capability of a worker, machine, work center, plant or organization to produce output per time period. Capacity required represents the system capability needed to make a given product mix (assuming technology, product specification, etc.). As a planning function, both capacity available and capacity required can be measured in the short term (capacity requirements plan), intermediate term (rough-cut capacity plan), and long term (resource requirements plan). Capacity control is the execution through the I/O

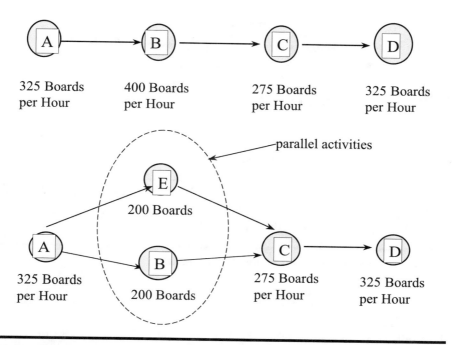

Figure 6.4 System Capacity is Determined by the Bottleneck (Adapted from Melnyk and Denzler, 1996)

control report of the short-term plan. Capacity can be classified as budgeted, dedicated, demonstrated, productive, protective, rated, safety, standing, or theoretical.

APICS Dictionary, 8th edition, 1995

When managers speak of capacity they may be talking about the short-term, medium-term, or long-term capacity of the firm. The decisions that are made in each time frame are dictated by the difficulty of changing or adjusting the capacity. In the short term, the bottleneck or constraint in the shop dictates the capacity. This constraint may be physical (e.g., a machine) or it may be labor (i.e., a shortage of skilled welders). This is shown in Figure 6.4. Capacity decisions in the short term usually involve whether to increase short-term capacity by adding overtime or how to maximize profit from the existing capacity. For example, if there is only enough capacity to make 100 units of a product in the short run, the firm makes the most money if it produces the most profitable units. In this situation a firm will turn work away because it cannot make the product.

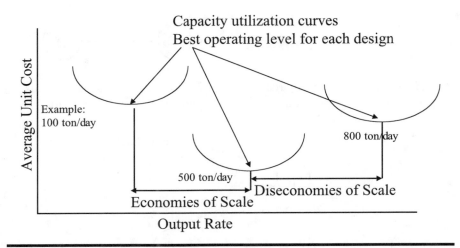

Figure 6.5 Economies & Diseconomies of Scale

The medium-term capacity decision is the number of workers to employ and the types of workers to have employed. This is where decisions are made about hiring or firing. The long-term decision is the decision about purchasing equipment or building new facilities.

A term that is often used when discussing capacity is *economy of scale*. It is important to understand that we have achieved economies of scale when the average unit size goes down as the capacity of the facility increases. This is illustrated in Figure 6.5 where the average unit cost for a ton of steel is reduced as the facility goes from an operating capacity of 100 tons/day to 500 tons/day. Another important concept is *diseconomies of scale*. This is also illustrated in Figure 6.5 where the cost of producing a ton of steel is increased when the plant grows from 500 tons/day to 800 ton/day. Economies of scale are gained by spreading the fixed costs over a larger volume. And, diseconomies of scale usually are the result of increased administrative costs, as it becomes more expensive to manage the larger facility.

The capacity utilization curve is also illustrated in Figure 6.5. The capacity utilization curve represents the amount of the existing capacity that is actually being used. The average unit cost of a product increases if the utilization is not within the planned region.

Economies of scale are achieved when the facility is being constructed or the equipment is being installed. It is due to the design of the equipment and facilities and the creation of an organization structure that can effectively operate these that firms achieve economies of scale. This decision determines what the average unit cost of production will be. The utilization of the capacity in the facility after it is constructed is determined by management. The capacity utilization also influences the average unit cost of the products produced, but the capacity utilization cannot change the position of the curve. Costs can increase whenever utilization is below or above the preferred operating range. If capacity utilization is low, costs increase because the fixed costs are spread out over fewer units. If capacity utilization is too high, then costs increase as the firm adjusts to the high demand. For example, the firm might have to work excessive overtime, or use expensive transportation sources to rush product to the customer.

Bottlenecks

The operating capacity of a system is determined by the bottleneck or the constraint in the system. A bottleneck is that part of the system where demand is greater than or equal to capacity. While the rest of the system has more capacity than the bottleneck, the system's output is limited by the output of the bottleneck.

Bottleneck—A facility, function, department, or resource whose capacity is less than the demand placed upon it. For example, a bottleneck machine or work center exists where jobs are processed at a slower rate than they are demanded.

APICS Dictionary, 9th edition, 1998

In Figure 6.4, there are two different production lines shown. In the top line there are no parallel activities and C has the lowest capacity (275 boards per hour) of all of the processes. So, C is the bottleneck. In the bottom part of Figure 6.4, C still has a capacity of 275 boards per hour, but there are two parallel activities (E and B), each of which has a capacity of 200 boards per hour. But, C is still the bottleneck, because it has the least amount of capacity. Since E and B perform the same process, they can operate in parallel and the process capacity is the sum of those machines which perform that process. In this case, the sum of the capacities of E and B is 400 boards per hour.

It is important in any operation to identify the location of the bottleneck. The bottleneck controls the flow of work through the firm, so the flow of work is easier to manage, if it can be controlled by managing the flow of work through the bottleneck.

Summary

This chapter identified 3 operating characteristics of a firm and discussed how these influence management of the firm. The desired customer lead time and the time required to produce the good or service dictates the company's decisions about stocking inventory and organizing its production facilities.

The selection of the production process and its layout is dictated in part by the volume of product needed and the flexibility needed in the production design. The level of capacity is another basic decision. This will dictate response times to customers and it will determine the volume of product the plant needs to be profitable. A final operating characteristic discussed is the presence of a physical bottleneck in the plant. This bottleneck controls the flow of work through the plant. The relevant issues about bottleneck and constraint management are discussed in greater detail in the next chapter, specifically relating to strategic selection of constraint locations.

7 Basics of Constraints Management

The strength of any chain is determined by its weakest link. That fundamental truth is at the root of the management concept called the theory of constraints. The implication in the management philosophy is that all systems will have at least one constraint and it is better to *manage* it rather than to constantly try to eliminate it. Once you strengthen the weakest link, some other link will become the weakest. In every system, regardless of its size or complexity, there will be a constraint. Realizing that the concept is not a theory but a reality, the names most often used today for this business philosophy are *constraints management* and *synchronous flow*.

Presentation to the market on a timely basis is the greatest opportunity for competitive advantage available to the manufacturing industries in the modern world. Be able to satisfy the market demand NOW, assuming you have good quality and competitive pricing, and you are likely to have all the business you can handle. This means that you must operate under a system that allows very short production cycle times so that your lead time to the customer is significantly better than that of the competition. *Lead time* is defined as the number of days (or hours) you can quote to your customer for delivery of a product or production order. Lead time is the total production cycle, which includes all the work-in-process currently in the system; the raw material purchase cycle; and any order backlog position that exists at that time. If it is ordered today, without pulling this order around any that are now in the pipeline, how long will it take to get it delivered to the customer?

The need to address this dilemma was the basis for development of constraints management. How do we operate a manufacturing system so that it will meet short cycle production demands without the need to maintain large stocks of finished goods inventory and, at the same time, is resistant to the effects of variability? The concept of constraints management offers a unique approach to this dilemma. The entire system is like a chain. The strength of a chain, like the strength of a manufacturing system, is dependent on the weakest link. You cannot get more through the system than the capacity of the weakest link or constraint of the system. By focusing on the performance of the system's constraint, rather than on the performance of each resource, the highest total system productivity is achieved with the available resources.

Protective Capacity

Synchronizing a manufacturing system involves selecting a constraint and operating the system with unbalanced capacity, also called *protective capacity*. The closer the system is to a balanced state, the more unstable it becomes. In other words, since a truly balanced capacity system is a practical impossibility, when there is an attempt to create balance, a condition called *the wandering bottleneck* occurs. Wherever "Murphy" last visited is the current bottleneck. The management/supervisory staff finds itself always reacting to the latest production problem. There is little opportunity to be proactive, because there is no way to predict where or to what degree normal variability will appear.

Let us be clear on the definition of a constraint. A *constraint* is anything that limits the degree to which an organization can satisfy its purpose. This is similar to the bottleneck definition given in Chapter 6, but the difference is that a constraint can be strategically located rather than allowing its location to be determined by chance. This will be explained in greater detail in the next section.

Types of Constraints

There are three types of constraints. Physical (logistical) constraints are the most obvious. They are resources within the system, which have a capacity that is equal to or less than the demand placed upon it. Physical constraints can be both internal (the capacity of a given resource) and external (the capacity of a supplier to provide the necessary raw material or even the market when manufacturing capacity exceeds the demand of the market). Policy (managerial) constraints are decrees or rules from the management staff that set limits on the system's performance in that they do not lead directly to achieving the goals and objectives of the system. Causing the system to emphasize efficiency or resource utilization is an example of policy constraint.

Paradigm (behavior) constraints are entrenched habits or assumptions of people in the system that "things must be done this way because they have always been done this way." Ironically, paradigm constraints often lead to policy constraints, which may lead to physical constraints. One thing is for sure: If the goal of the organization is to continually increase value added, and its actual value added is something less than infinite, then a constraint exists. Every system, without exception, has a constraint.

The synchronous manufacturing approach requires that the constraint be clearly visible. The capacity of the constraint versus that of the nonconstraints must be significant enough so that normal production problems (such as machine malfunctions and operator absenteeism) do not disrupt the product flow. There must be "protective capacity" at all the nonconstraints. The capacity of these resources must be sufficient to absorb the normal system variability without starving the constraint of work. Also, there must be a sufficient buffer of work preceding the constraint to act as a shock absorber for these normal fluctuations in production. In fact, there is a dynamic relationship between productive capacity (the capacity of the constraint which is the capacity of the system), protective capacity (the capacity of the nonconstraints which must be greater than that of the constraint), and inventory (the amount of buffer within the system to protect the constraint). The greater the differential between protective and productive capacities, the smaller the buffer inventory needs to be. Conversely, the closer this differential is to being even, the greater the inventory must be to protect the constraint from starvation. Variability is the compounding factor in this relationship. The greater the system variability (normal statistical fluctuations), the more protective capacity and/or inventory is needed to maintain stability. Keep in mind that anytime the constraint is starved, production is lost for the whole system.

Drum–Buffer–Rope

The concept just described is called Drum–Buffer–Rope (DBR) scheduling (see Figure 7.1). The constraint is the drumbeat of the system. Ideally, the constraint never stops working. It is always producing or setting up to be producing products that the system can ship. It is protected by a buffer of work-in-process to assure that it always has work to do. As variability occurs at any of the resources feeding the constraint, the buffer is depleted. By definition, the nonconstraints have a greater capacity than the constraint, so when the problem is corrected, the nonconstraint has the time to catch up before the constraint depletes the buffer. When the buffer is replenished to its specified

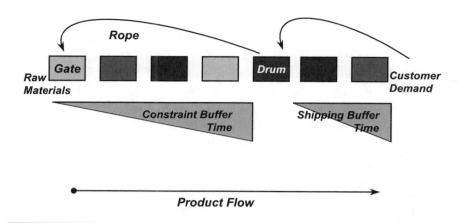

Figure 7.1 Synchronized Flow

level, the nonconstraint stops working on that operation to avoid unnecessary inventory build up. In fact, the amount of raw material released into the system is controlled by the rope based on the consumption of that raw material at the constraint. The rope is like a signal from the constraint indicating the amount of raw material to be released.

There are several types of buffers. The constraint buffer protects the constraint's ability to meet its schedule. The shipping buffer protects due date integrity, given that due date performance is the first constraint imposed on the system. The assembly buffer prevents constraint parts from waiting on nonconstraint parts at assembly. A raw material buffer will protect the ability to meet the release schedule against nonperformance of the raw material suppliers. The size of the buffer is expressed in terms of time. The level of inventory in the buffer is converted into time by determining how long it would take the bottleneck or constraint to produce that amount of inventory. For example, if the shipping buffer has 20 products in it and each of these requires 30 minutes of processing on the bottleneck, then the shipping buffer is a 10-hour buffer.

The concept of DBR contains several basic algorithms (see Figure 7.2), which are:

- The customer due date minus the shipping buffer equals the constraint's due date.
- The constraint's due date minus the constraint processing and setup times equals the constraint start date.
- The constraint start date minus the constraint buffer equals the material release date.

- due date – shipping buffer = constraint's due date
- constraint's due date – (constraint processing + setup time) = constraint start date
- constraint start date – constraint buffer = material release date

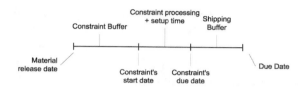

Figure 7.2   Basic DBR Algorithm

The Prerequisites

↙ Define the system and its purpose.

↙ Decide how to measure it.

The Focusing Steps

❶ Identify the system's constraint(s).

❷ Decide how to exploit the system's constraint(s).

❸ Subordinate everything else to the above decisions.

❹ Elevate the system's constraint(s).

❺ Do not allow inertia to become the system's constraint. When a constraint is broken, return to step one.

Figure 7.3   The Focusing Steps

The Five Focusing Steps

There are five focusing steps to applying this concept to a manufacturing system (see Figure 7.3). It is impossible to "focus" on everything (as is the normal strategy) so we should decide what resources are truly determining the capacity of the system and concentrate our efforts there. Just as the weakest link of a chain determines the strength of the chain, the system's constraint determines the capacity of the system. Every system has a constraint, just as every chain has only one weakest link.

Before applying the five focusing steps, it is important to define the system or the scope of the process to be synchronized. The system could be a plant, a value stream within a plant, or a supply chain composed of several plants including suppliers and the customer. In any case, it is important to first decide what the boundaries of the system to be synchronized are. Next, we must describe the purpose of the system and decide how to measure it. As stated earlier, we must know just what we are seeking and we must set up measurements to achieve those results. It cannot be overemphasized that the measurements will determine the performance, so we had better be monitoring the important things and not just the traditional cost accounting factors such as efficiency and utilization.

Having defined the system and its purpose including the measurements to be achieved, the first of the focusing steps is to *identify* the constraint. Find it or pick it. In other words, determine where the constraint naturally exists or pick a place where you strategically want the constraint to be. This will be the focal point of the entire synchronized system, so it is an important decision. It is important to note that, in the context of synchronous manufacturing, a constraint is actually a good thing as long as we have correctly identified it and we use its performance to manage the rest of the system.

The second of the focusing steps is to *exploit* the constraint. Exploitation means assuring that the constraint is working *only* on products that will ship and for which the system will receive payment. It means assuring that only first quality parts are allowed to pass through the constraint. We would not want to waste any of this valuable constraint time on products that will not be shippable. Also, exploit means working the constraint during every available hour of the work day. Any time lost at the constraint is time lost for the system.

The third focusing step is to *subordinate* everything else in the system to the decision on where the constraint is to be located. Subordination is the most difficult of the synchronous flow methods. Every other management decision including the release of raw materials into the system must be based on the consumption at the constraint. Releasing raw materials at a faster rate just to keep some resource busy will only add unneeded inventory in the system, which will lengthen the customer lead times. If only we worried as much about idle inventory as we do about idle workers, our plants would be much more responsive to the market.

Having completed the first three focusing steps, the only way to get more out of the system is to *elevate* the constraint. That means to add more capacity at this critical resource. This could mean adding equipment or people, or it could mean offloading some of the constraint duties to other resources. The net result is more capacity at the constraint, which always results in more capacity for the entire system. However, if we elevate the constraint, it is possible that it will be broken. That is, the constraint we identified may no longer be the constraint. Whenever this happens, the logistical constraint will move to some other location in the system, internally as another resource or externally as at a supplier or in the market itself. We must then *return to step one* and rethink the entire system identifying the new constraint. It is important here to watch out for inertia in the form of policy or paradigm constraints that may not have been issues when the constraint was in its original location, but now may well be. Because everything in the system must be subordinated to the performance of the constraint, we often create rules to assure that this happens. When the constraint moves, those rules are no longer valid, so it is imperative that we review the policies (both formal and informal) that have been developed. Where they are no longer needed, they must be voided.

Summary

The process of synchronous manufacturing is largely intuitive, but is usually contrary to the accepted practices of management, particularly if management decisions are made based on cost accounting practices. If we can look beyond the policies and rules that govern our organizations, the ideas of synchronizing the flow of material through the system based on consumption of an identified constraint seem very intuitive. However, most companies do not look at their systems in this manner. The constraints management business philosophy is the process by which an organization can take a commonsense look at the whole system and achieve results that are uncommon by today's standards.

The constraint management is truly not limited to manufacturing organizations. Service companies, distribution systems, and even not-for-profit organizations are using the same principles previously used only in manufacturing to help them reach their goals. Every system, no matter what its process or its ultimate objective, has a constraint and its performance can be maximized by applying the principles of constraints management. This is a process of continuous improvement, so it is being utilized in all types of organizations as a methodology to continue raising the bar of performance.

Constraint management techniques are proven to offer solutions to the problems facing the world's manufacturing base. The commonsense approaches of this philosophy allow higher productivity, lower work-in-process levels, and faster processing times which lead to shorter customer lead times. Application to the entire supply chain will offer increase advantages by providing the tools of communication and synchronization that are critical to optimum performance.

CUSTOMER ORDER CYCLE

T his section examines the customer order cycle, which is the flow of work that begins with the customer's order and finishes when the product is delivered to the customer. Managing this cycle is a vital ingredient in managing the supply chain. Chapter 8 will discuss the different techniques used by companies to manage their linkage with their customers. Chapter 9 will discuss how companies use information from customers and suppliers to transform their inputs into products that their customers want. Chapter 10 will discuss how companies manage their suppliers.

8 | The Customer Linkage

This chapter will examine how firms manage their linkage with their customers. This is a vital link and the ability to manage it better than your competitor will give your firm a competitive advantage in the market. What makes your firm's management of this linkage better than your competitor's management of this linkage can be measured using the quality management measurements discussed earlier.

Forecasting

There are products that the customer wants delivered immediately. When you go to a hardware store you do not want to place an order to buy a box of nails or a hammer. You expect to be able to carry them home. It is still necessary to estimate final demand for these consumer items. This process of estimating final consumer demand is called *forecasting*. In a well-run supply chain only the final consumer demand is forecasted. All of the components that go into the product do not have to be forecast, because they can be calculated based on the forecast for consumer demand. For example, if there is a forecast by a retail store that it will sell 100 24" bikes in June, the bike manufacturer knows that they will need 100 front wheels and 100 rear wheels. These components do not have to be forecast.

Eliminating the need to forecast by supplying real information shortens the product lead time. A firm does not use its capacity to produce products that may not be needed; instead it uses its capacity to meet the actual demand. This does not mean that there is no inventory in the system. For example, a firm may receive information from Wal-Mart about how many saw blades were sold in a week. The firm then schedules production. But, Wal-Mart may carry a two-week supply of inventory to protect itself from surges in demand and to allow for lead time. So, the firm is actually replacing inventory when it builds.

95

As communication improves throughout the extended supply chain, the importance of forecasting future demand may be reduced. For example, electronic data interchange (EDI) allows the customer's demand data to be transmitted quickly to the supplier. When the members of the supply chain use the same communication conventions they can take this customer information and use it for planning. If all customers have EDI and communicate with the supplier there is no need for forecasting for suppliers who do not serve the retail customer. Another technology which is reducing the need for forecasting is the Internet, because it allows firms to communicate rapidly with each other. Firms may use the Internet to share information about their demand with each other. However, these technologies have not yet eliminated the need to forecast for most members of the supply chain. Because most firms have customers who do not communicate with them using either EDI or the Internet, forecasting still remains an important issue.

The forecast allows the company to plan and it allows the company to coordinate internally. It also allows the entire chain to coordinate among themselves to serve the customer. For example, if marketing produces a forecast to sell 100 units, operations will coordinate by scheduling to build 100 units and purchasing will coordinate by obtaining the materials to build 100 units.

Forecasting is the prediction of future demand. Different forecasts are made for different purposes. Long-range forecasts are used to determine when and where to build facilities and how large the facilities should be. Medium-range forecasts are used to determine the staffing levels of facilities currently being operated by a company. Short-range forecasts, for as short a period as a week extending out to about 6 months, are used to estimate customer demand.

It is important to remember that every forecast will be wrong. Forecasts are always a prediction of the future. It is random luck if demand actually equals the forecast. As shown in Figure 8.1, there is a distribution of probable demand around the forecast point. So, a good forecast includes not only an expected amount, but also the probable range as shown by the distribution. The reason we make the forecast is to coordinate between all the functions and firms in the supply chain. To help coordinate the supply chain the forecast needs to be reasonably accurate. In the next sections we will consider some simple forecasting methods and some simple methods to evaluate the accuracy of the forecast.

The forecasting calculation has been greatly simplified in recent years by the introduction of PCs with user-friendly forecasting software. There is a large number of forecasting methods and one of the major tasks for the forecaster is to select the most appropriate technique for making the forecast.

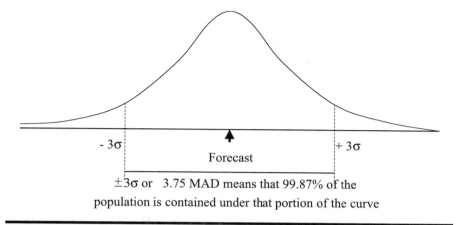

$\pm 3\sigma$ or   3.75 MAD means that 99.87% of the
population is contained under that portion of the curve

Figure 8.1   Distribution of Demand Around Forecast

Common Forecasting Methods

There are many types of forecasting methods in common use. The selection of the method to use depends on the industry of a firm and the length of time that is being forecasted. The simplest forecasting technique is the naive method. It simply states that demand this next period will be what demand was the last period. This is not a useful method for most industries. It is however effective in some industries, such as the fast food industry. In that industry the store manager predicts sales for the next Monday based on the sales of the prior Monday and orders an appropriate amount of food and schedules the necessary staff.

> **Forecast**—An estimate of future demand. A forecast can be determined by mathematical means using historical data, it can be created subjectively by using estimates from informal sources, or it can represent a combination of both techniques.
>
> *APICS Dictionary,* 8th edition, 1995

A common set of forecasting techniques are called *time-series forecasts.* A *time series* is any group of data that is arranged in sequence according to the time it was gathered. For example, the monthly demand for a product is time-series data. There is a large number of time-series forecasting methods. We will examine only three of them.

Time-Series Forecasting

The *moving average* is a simple times-series method. It is just the average of the demand for a set of the most recent periods. For example, if demand for the last 4 weeks was 100, 120, 130, and 120 for weeks 1 to 4, respectively, and if the number of periods to be averaged is n, then $n = 4$ and the moving average is calculated as:

$$MA = (d_1 + d_2 + d_3 + d_4)/n = (100 + 120 + 130 + 120)/4 = 117.5 \cong 118$$

In the example above, the demand in week 1 was 100, in week 2 it was 120, in week 3 it was 130, and in week 4 it was 120. The average demand was rounded up to 118. So, using this method the forecast for period 5 is 118.

Continuing with this example, if demand in week 5 was 140 the moving average is calculated as:

$$MA = (d_2 + d_3 + d_4 + d_5)/4 = (120 + 130 + 120 + 140)/4 = 127.5 \cong 128$$

So the forecast for period 6 is 128.

Time series—A set of data that is distributed over time, such as demand data in monthly time period occurrences.

APICS Dictionary, 8th edition, 1995

Time series analysis—Analysis of any variable classified by time in which the values of the variable are functions of the time periods.

APICS Dictionary, 8th edition, 1995

Note in this last example that the demand from the period that occurred the longest time ago was discarded. The latest available demand was included in the calculation. Examining these two equations, it is possible to see that if *n* is reduced in size, the forecast resembles the latest demands. If $n = 1$, then the moving average becomes the naive forecasting method. By keeping *n* small, the method is very responsive to changes in demand. When *n* is increased in size, then the moving average responds very slowly to increases or decreases in demand.

Moving average—An arithmetic average of a certain number (*n*) of the most recent observations. As each new observation is added, the oldest observation is dropped. The value of *n* (the number of periods to use for the average) reflects responsiveness versus stability . . .

APICS Dictionary, 8th edition, 1995

Weighted moving average—An averaging technique in which the data to be averaged are not uniformly weighted but are given values according to their importance.

APICS Dictionary, 8th edition, 1995

A second time-series forecasting method is the *weighted moving average.* Instead of simply taking the average of a set number of periods, this method weights the more recent periods heavier than the older periods.

Continuing to use the example above, where demand for the last 4 weeks was 100, 120, 130, and 120 for weeks 1 to 4, respectively, and with weights (w_1, w_2, w_3, and w_4) of .1, .2, .3, and .4 for weeks 1 to 4, respectively, the weighted moving average is:

$$WMA = w_1 d_1 + w_2 d_2 + w_3 d_3 + w_4 d_4 = .1 * (100) + .2 * (120) + .3 * (130) + .4 * (120) = 121$$

In this example, the weighted average demand was 121. So, the forecast for period 5 using this method is 121. Continuing with this example, when we are finished with period 5 and are forecasting demand for period 6, we take the actual demand from week 5 into the formula and drop the oldest actual demand out of the formula. If week 5's actual demand was 140, then the weighted moving average is calculated as:

$$WMA = w_2 d_2 + w_3 d_3 + w_4 d_4 + w_5 d_5 = .1 * (120) + .2 * (130) + .3 * (120) + .4 * (140) = 130$$

It is important to note in this last example that the demand from the period that occurred the longest time ago was discarded. The latest available demand was included in the calculation.

A third time-series forecast method is *exponential smoothing.* This exponential smoothing method is just a weighted moving average that includes Greek letters in the formula. But, because it always uses the prior periods forecast in the forecast for the next period, it does not completely ignore any data from the past. A major advantage of the method is that it requires very few calculations.

Exponential smoothing—A type of weighted moving average forecasting technique in which past observations are geometrically discounted according to their age. The heaviest weight is assigned to the most recent data. The smoothing is termed exponential because data points are weighted in accordance with an exponential function of their age. The technique makes use of a smoothing constant to apply to the difference between the most recent forecast and the critical sales data, thus avoiding the necessity of carrying historical sales data. The approach can be used for data that exhibit no trend or seasonal patterns. Higher order exponential smoothing models can be used for data with either (or both) trend and seasonality.

APICS Dictionary, 9th edition, 1998

Continuing to use the example above, where demand for the last 4 weeks was 100, 120, 130, and 120 for weeks 1 to 4, respectively, we will forecast demand in week 5. To do this we must have the forecast for period 4; to get the forecast for period 4, we need the forecast for period 3. This continues until period 1, when we recognize that we need to initiate the method by creating a forecast for period 1 using a different method. One common technique to obtain the first week's forecast is to say the forecast for period 1 was equal to actual demand for period 1 (i.e., we set the forecast for week 1 equal to 100 since actual demand for week 1 was 100). The formula for the exponential moving average is:

$$F_{t+1} = \alpha d_t + (1 - \alpha)F_t \ where \ 0 < \alpha < 1$$

This formula states that the forecast for the next period (F_{t+1}) is equal to a weight (α) times the demand this period (d_t) plus another weight ($1 - \alpha$) times the forecast (F_t) for this period. Note that α must stay between 0 and 1. If $\alpha = 1$, then the forecast for next period becomes the demand for this period, which is the naive forecast model discussed earlier. If $\alpha = 0$ then the forecast for next period is the forecast for this period, which means the forecast would never change. By looking at the equation you can see that the larger the value we choose for α, the more responsive the model will be to changes in demand, because as α approaches 1, the term $(1 - \alpha)$ approaches 0 so that the forecast for the next period does not consider the forecast for this period. To simplify the calculations the terms of the exponential smoothing equation can be rearranged as:

$$F_{t+1} = F_t + \alpha(d_t - F_t) \ where \ 0 < \alpha < 1$$

This simplified version of the exponential smoothing equation is often faster to compute on a calculator and is easier to understand intuitively. This equation states that the forecast for the next period (F_{t+1}) is equal to the forecast for this period (F_t) plus a weight (α) times the error in this period's forecast ($d_t - F_t$).

Using the example from above where the demand in week 1 was 100, in week 2 it was 120, in week 3 it was 130, and in week 4 it was 120, we set the first week's forecast equal to demand for the first week (i.e., $F_1 = 100$). If we select $\alpha = .2$, then the forecasts for weeks 2 to 4 are calculated as:

$$F_2 = F_1 + \alpha(d_1 - F_{1t}) = 100 + .2(100 - 100) = 100$$

Using the forecast of $F_2 = 100$, we calculate F_3 as:

$$F_3 = F_2 + \alpha(d_2 - F_2) = 100 + .2(120 - 100) = 104$$

In a similar manner we calculate F_4 and F_5 as:

$$F_4 = F_3 + \alpha(d_3 - F_3) = 104 + .2(130 - 104) = 109.2$$

$$F_5 = F_4 - \alpha(d_4 - F_4) = 109.2 + .2(120 - 109.2) = 111.36$$

Continuing with this example, if the actual demand in week 5 was 140 the exponential smoothing forecast for week 6 is calculated as:

$$F_6 = F_5 - \alpha(d_5 - F_5) = 111.36 + .2(140 - 111.36) = 117.1$$

All of the time-series methods are used for short-term forecasts. They are used to calculate the demand for the next period.

Forecasting with Regression Models

A third type of forecasting model can be used for long-range forecasts. These are *linear regression models.* This type of model assumes that the time-series data is linear (i.e., it can be represented by a straight line). When this is the case then it is possible to forecast demand as:

Forecast = Constant + Slope*(number of periods in the future)

Regression analysis—A statistical technique for determining the best mathematical expression describing the functional relationship between one response and one or more independent variables.

APICS Dictionary, 9th edition, 1998

> **Least-squares method**—A method of curve fitting that selects a line of best fit through a plot of data to minimize the sum of squares of the deviations of the given points from the line.
>
> *APICS Dictionary,* 9th edition, 1998

The value of the constant and the slope in the equation are calculated using the linear regression formulas for the terms. These terms can be estimated using the Regression Function in Excel or another spreadsheet. Or, they can be calculated by hand using the formulas found in any statistics text. The method that is used to calculate them is the least squares estimate. This method minimizes the sum of the square errors between the actual value and the predicted value.

The slope coefficient (b) and the intercept (a) are calculated using the formulas below.

$$b = \frac{\Sigma(X_i - \overline{X})(Y_i - \overline{Y})}{\Sigma(X_i - \overline{X})^2} \qquad a = \overline{Y} - b\overline{X}$$

This formula says that the slope coefficient is the sum of the differences between the predictor variable (X) and its average $(\overline{X})$ and the dependent variable (Y) and its average $(\overline{Y})$ divided by the sum of the squares of the difference between each predictor variable and the average of the predictor variables. The intercept is equal to the average of the dependent variables minus the slope times the average of the independent or predictor variable.

Using the example from above where the actual demand was 100, 120, 130, and 120 for weeks 1 to 4, respectively, we first state what our model is. In this case we are predicting that demand change is due to the passage of time. So, the predictor variable is t, the length of time, and the dependent variable is the actual demand. The average of the time periods 1 to 4 is 2.5. The average of the actual demand is 117. The regression equation with these values entered into it is:

$$b = \frac{\Sigma(X_i - 2.5)(Y_i - 117)}{\Sigma(X_i - 2.5)^2} \qquad a = 117 - b*2.5$$

Substituting the values in for X and Y, we find that b = 7 and a = 100. So, the regression equation for our forecast is:

$$\text{Forecast}_t = 100 + 7t$$

In this equation demand for t periods in the future is equal to 100+7 times the number of periods. So, for period 5 the forecast is:

$$\text{Forecast}_5 = 100 + 7*5 = 135$$

And for period 6, the forecast demand is:

$$\text{Forecast}_6 = 100 + 7*6 = 142$$

This equation can be used to forecast into the future as far as needed. But, there is danger in using this forecast equation for anything besides a short-range forecast. This method is still predicting the future based on the past. If it is used to predict too far into the future, it is possible that the technology or customer demand could change significantly enough that past data will not help predict the future. For example, a new type of disk drive may be introduced and demand drops for the old model of desktop PC disk drive.

Forecast Error

At this point you have probably noticed that all the different forecasts made with different forecasting models gave different forecasts. Which one should you use? Usually managers select the forecasting model that they will use to make forecasts using empirical evidence of how well it has made forecasts in the past. To evaluate the accuracy of each forecasting model, managers measure the forecast error. The manager will then select the forecast model that provides the least amount of error. The forecast error is the difference between actual demand and the forecast of the demand.

Forecast error—The difference between actual demand and forecast demand, stated as an absolute value or as a percentage.
APICS Dictionary, 9th edition, 1998

Mean absolute deviation (MAD)—The average of the absolute values of the deviations of observed values from some expected value. MAD can be calculated based on observations and the arithmetic mean of those observations. An alternative is to calculate absolute deviations of actual sales data minus forecast data. These data can be averaged in the usual arithmetic way or with exponential smoothing.
APICS Dictionary, 9th edition, 1998

Table 8.1 Calculation of Mean Absolute Deviation (MAD)

Period	Actual Demand	Forecast	Forecast Error	Absolute Error
1	100	90	10	10
2	120	110	10	10
3	130	128	2	2
4	120	130	−10	10
5	140	120	20	20
			MAD =	10.4

A common measure of forecast error is the mean absolute deviation (MAD). The MAD is easily calculated and is easily interpreted. The MAD is calculated by adding the absolute value of the forecast errors each period (| Demand − Forecast |) and then taking the average of this total. This is illustrated in Table 8.1.

For clarity, this example first calculated the forecast error. It then took the absolute value of that forecast error, and, finally, the average of the absolute value was calculated. The MAD gives us a measure of the distribution of the forecast error, so we can estimate a minimum estimate and a maximum estimate for demand. For example, if our forecast model forecast 130 for period 6, we would know that that if the actual demand was within ± 1 MAD from the point estimate of demand (i.e., 130) it could be as high as 140 and as low as 120. Statisticians have calculated that 1 MAD = 0.80 standard deviations, so that 3.75 MADs is equivalent to 3.00 standard deviations (see Melnyk and Denzler, 1996, for more information). Because ± 3.0 standard deviations includes 99.87% of the population, when forecasters provide a forecast and the value of one MAD, they allow the user to quickly determine the accuracy of the forecast. This is illustrated in Figure 3.2. In this example, ± 3.0 standard deviations means that there is a 99.87% chance that the actual demand will be between ± 39 of the forecast of 130. Or, the actual demand has a 99.87% chance of being between 91 and 169. This is calculated as 130 ± (3.75)*(10.4).

By sharing the MAD with the manager, the forecaster reminds everyone that the basic law of forecasts is that they will be wrong. By selecting a different model, the forecaster may be able to reduce the MAD dramatically, but it cannot be eliminated.

Another measure of forecast error is mean squared error (MSE), which is the average of the square of total forecast errors for a sample. This approximates the variance. The formula to calculate the MSE is:

$$MSE = \sum_{t-1}^{n} (demand_t - forecast_t)^2/(n-1)$$

The MSE for the example above is calculated below in Table 8.2.

Table 8.2   Calculation of Mean Squared Error (MSE)

Period	Actual Demand	Forecast	Forecast Error	Squared Error
1	100	90	10	100
2	120	110	10	100
3	130	128	2	4
4	120	130	−10	100
5	140	120	20	400
			MSE =	176

Forecast Bias

Bias—A consistent deviation from the mean in one direction (high or low). A normal property of a good forecast is that it is not biased.

APICS Dictionary, 9th edition, 1998

It is important to distinguish systematic error from random error. To do this, the forecaster measures the amount of bias in the forecast. The *APICS Dictionary* defines bias as "A consistent deviation from the mean in one direction (high or low). A normal property of a good forecast is that it is not biased." The usual measure of bias is the mean forecast error (MFE) which is the mean of the difference between the actual demand and the forecast demand. The formula for the MFE is:

$$MFE = \frac{\sum_{t=1}^{n} (D_t - F_t)}{n}$$

Remember, no forecast will be correct. If there is just random error, then about 50% of the time the forecasts will be higher than the actual demand and about 50% of the time the forecasts will be lower than actual demand. If the MFE is close to zero, then the forecast model is not biased. If the MFE is negative then the forecasts are biased on the high side of the distribution (i.e., the forecast is too large). If the MFE is positive then the forecasts are biased on the low side of the distribution. In Table 8.3, the MFE is 6.4. This means that if there is any bias, it is that that the forecast is consistently low.

Table 8.3   Calculation of Mean Forecast Error (MFE)

Period	Actual Demand	Forecast	Forecast Error
1	100	90	10
2	120	110	10
3	130	128	2
4	120	130	−10
5	140	120	20
		MFE =	6.4

Tracking Signal

Tracking Signal—The ratio of the cumulative algebraic sum of the deviations between the forecasts and the actual values to the mean absolute deviation. Used to signal when the validity of the forecasting model might be in doubt.

APICS Dictionary, 9th edition, 1998

There are many other measures of forecast error, but the only other measure we will examine here is called the *tracking signal.* The *tracking signal* is the sum of the forecast errors divided by the mean absolute deviation (i.e., the MAD). The tracking signal is used to signal the forecaster when the validity of the forecasting model is doubtful. The tracking signal is really a record of each product's cumulative forecast error and using it avoids the need to maintain a control chart for each item in stock. The formula is:

$$Tracking\ Signal = \frac{\sum_{t=1}^{n}(D_t - F_t)}{MAD}$$

The tracking signal is recalculated each period and compared with a pre-set value which is usually somewhere between ±3 standard deviations and ±8 standard deviations. Whenever the tracking signal calculation is larger than this preset limit, the forecaster looks for a problem in the forecast. An example of the use of the tracking signal is shown in Table 8.4. Each period the MAD is calculated and the sum of the forecast error to that period is calculated and divided by the MAD for that period. The result is the tracking signal, which is compared to the critical value for the tracking signal. If the calculated tracking signal is smaller than the critical value, then the forecaster assumes that nothing is wrong with the forecast. If the tracking signal is larger than the critical value, then the forecaster investigates to find the potential problem in the forecast.

For example, in period 2 of Table 8.4, the forecast error is 10. The sum of the forecast errors to that point is 20 (period 1 + period 2). The MAD for the first two periods was 10. So, dividing the sum of the forecast error (20) by the MAD (10) gives a tracking signal of 2. This is less than the tracking signal critical value, so no problem is indicated in the forecast.

Table 8.4   Tracking Signal Calculations

Period	Actual Demand	Forecast	Forecast Error	Absolute Error	MAD	Tracking Signal
1	100	90	10	10		
2	120	110	10	10	10	2
3	130	128	2	2	7.33	3
4	120	130	−10	10	8	1.5
5	140	120	20	20	10.4	3.08

Tracking Signal Critical Value = 4

Demand Management

In the section above, forecasting was viewed as a method of predicting demand from customers. This section examines demand management which is a term applied to all those methods used to influence customer demand. Examples include marketing promotions, advertising, customization, and sales.

As stated above, demand management is an attempt to influence the customer. The operations manager does not do this by taking over the marketing function. Rather, the operations manager becomes involved with demand management by determining (often in conjunction with marketing) what capabilities are necessary to influence customer demand and then creating those capabilities within the operations system. Examples of this might be to develop the capability of providing special packaging or of customizing the product or the order. Another common example is to develop the capability of delivering orders faster than the competition.

Demand management—The function of recognizing all demands for products and services to support the marketplace. It involves doing what is required to help make the demand happen and prioritizing demand when supply is lacking. Proper demand management facilitates the planning and use of resources for profitable business results. It encompasses the activities of forecasting, order entry, order promising, and determining branch warehouse requirements, interplant orders, and service parts requirements.

APICS Dictionary, 9th edition, 1998

Demand management can usually be effective only if there is some cross-functional means of coordinating efforts throughout the organization. This might be done by a product team or by a team of executives. The team is responsible for setting due dates, prices, and inventory levels and for allocating capacity. These are coordinated with the activities of other functions such as marketing.

Many of the demand management decisions made by a firm are implemented by the department responsible for order entry. It is this department which is responsible for communication with the customer. It is this department which quotes the prices and promises delivery dates to the customer. If demand is decreasing, management has the option of reducing the price or reducing the lead time to stimulate demand. Likewise, if demand is increasing, management can seek to control it either by increasing the lead time or by raising the prices.

Part of the demand management decision was discussed earlier when the firm selected its approach to the market. Is the firm a make-to-stock firm (MTS)? Or, is it an assemble-to-order (ATO), make-to-order (MTO), or engineer-to-order (ETO) firm? These organizational decisions influence the amount of risk the producer faces and the speed at which the product can be supplied to the customer.

Another part of the demand management decision is made by the planning department. The planning department can maintain some slack in the schedule of orders accepted so that order entry can accept rush orders from preferred customers.

9 Design and Management of the Transformation

Every firm belongs to at least one supply chain. Most firms belong to several supply chains. The question for each firm is how to design and manage the portions of the transformation processes that are its responsibility. For the supply chain to successfully meet the needs of the ultimate customer, the individual firms must design and manage their processes so that they add value to the ultimate customer. As shown in Figure 9.1 this could be difficult, because the firm may be a member of a supply chain that serves different sets of ultimate customers.

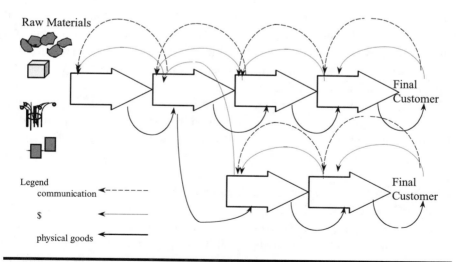

Figure 9.1   Multiple Supply Chains

Review of Volume and Layout Decisions

As discussed earlier in the book, the major determinant of the shop floor layout is the volume of product that the firm expects to produce. Firms can increase the total volume that is produced at one facility by being members of more than one supply chain. It is feasible for firms to do this when the products they are producing for each supply chain are similar.

For example, in the consumer goods industries the retail store is often the controlling member of the supply chain. In today's retail industry, the retail store must have the right product on the shelf at the right time. Many of these products are household brands which consumers expect to find in almost any retail store they visit. The factories supplying these products are then members of supply chains for competing retail stores. This increases the volume for the factory, but the factory may have to coordinate with the distribution systems of the different retail chains in different ways to meet the different requirements of these various retail stores.

Volume, Layout, and Competitive Factors

All supply chains exist to serve a customer. The ability of a supply chain to compete on the basis of one or more of the four competitive factors (i.e., cost, quality, delivery, and flexibility) is controlled by the ability of the supply chain members to compete on those competitive factors. A firm's position on each of these competitive factors is determined by decisions it makes in the eight manufacturing decision categories of capacity, facilities, technology, vertical integration, workforce, quality, production planning/materials control, and organization. The first four (i.e., capacity, facilities, technology, vertical integration) are structural decisions and the last four are infrastructure decisions. The structural decisions are difficult to change quickly. They are long-term investments that once made are difficult to change. So, the layout choice is often made within the constraints of these structural decisions. The layout decision, in turn, determines the firm's ability to deliver these competitive factors.

The layout of a firm's facilities determines a significant portion of how competitive the firm is. The layout determines both the ability of the firm to respond to change quickly and the material handling costs. Balancing the material handling costs are the costs of capital. A product layout requires more capital investment than a process layout. But, a product layout allows for faster response and for higher volume at a lower average unit cost.

Manufacturing Planning and Control Systems

The manufacturing planning and control system (MPCS) or, as it is sometimes referred to, the production planning and control system is critical to a firm's success. This system includes many management activities such as planning capacity, materials, monitoring equipment utilization, maintaining inventories, scheduling, setting customer due dates, and supplying information to the other functions.

The MPCS system involves the manager in two of the typical activities of management—planning and control. The MPCS is an integrated system that both creates plans and controls activities to ensure that they adhere to the plan.

Control system—A system that has as its primary function the collection and analysis of feedback from a given set of functions for the purpose of controlling the functions. Control may be implemented by monitoring or systematically modifying parameters or policies used in those functions, or by preparing control reports that initiate useful action with respect to significant deviations and exceptions.

APICS Dictionary, 9th edition, 1998

The major activities in the MPCS are illustrated in Figure 9.2. This is a hierarchical framework. The top level decisions are made first and then the increasingly detailed decisions of the lower levels are made. Note that the appropriate capacity management technique is used at the same time as each level of the plan is developed.

The production planning which is the first activity shown in Figure 9.2 is sometimes called *aggregate production planning, aggregate planning,* or *resource planning.* The term *aggregate* is sometimes used because at this level individual units or parts are not planned. Rather, production for a family of end items is planned. This is usually over a longer time horizon. The purpose of this plan is to adjust capacity to match expected demand and to have available the other resources that are needed. One of the major purposes of the production planning function is to coordinate not only the activities within operations, but all of the functions in the firm with the business plan. This means that marketing, production, engineering, and finance will all have input into the plan. The major concerns of production will be to ensure that

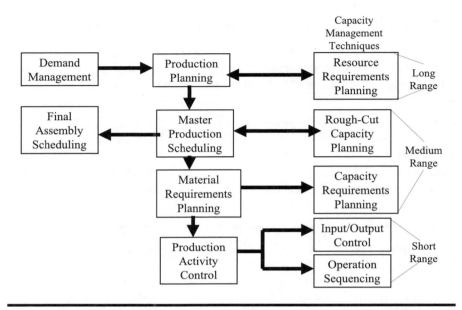

Figure 9.2   Capacity Decisions over the Time Horizon (Based on Blackstone, 1989, p. 43)

it has adequate capacity to meet the production objectives of the plan. It is for this reason that resource requirements planning or resource planning is conducted in conjunction with the development of the production plan.

Aggregate planning—The process of developing, reviewing, analyzing, and maintaining aggregate plans. Aggregate planning usually includes total sales, total production, targeted inventory, and targeted customer backlog of families of products. The production plan is the result of the aggregate planning process.

APICS Dictionary, 9th edition, 1998

Resource planning—Capacity planning conducted at the business plan level. The process of establishing, measuring, and adjusting limits or levels of long-range capacity. Resource planning is normally based on the production plan but may be driven by higher level plans beyond the time horizon for the production plan, e.g., the business plan. It addresses those resources that take long periods of time to acquire. Resource planning decisions always require top management approval.

APICS Dictionary, 9th edition, 1998

Production planning—The function of setting the overall level of manufacturing output (production plan) and other activities to best satisfy the current planned levels of sales (sales plan or forecasts), while meeting general business objectives of profitability, productivity, competitive customer lead times, etc., as expressed in the overall business plan. The sales and production capabilities are compared, and a business strategy that includes a sales plan, a production plan, budgets, pro forma financial statements, and supporting plans for materials and workforce requirements, etc., is developed. One of its primary purposes is to establish production rates that will achieve management's objective of satisfying customer demand, by maintaining, raising, or lowering inventories or backlogs, while usually attempting to keep the workforce relatively stable. Because this plan affects many company functions, it is normally prepared with information from marketing and coordinated with the functions of manufacturing, engineering, finance, materials, etc.

APICS Dictionary, 9th edition, 1998

The purpose of the plan is to ensure that production can meet management's objective of satisfying customer demand. One technique to do this in addition to obtaining more resources is to maintain, raise, or lower inventories or backlogs. This requires management to balance the risks of inventory obsolescence with the costs and disruptions of hiring and firing the work force. The customer demand information is an output of the demand management activity and, as discussed earlier, is a major input into the production planning process.

Demand management—The function of recognizing all demands for products and services to support the marketplace. It involves doing what is required to help make the demand happen and prioritizing demand when supply is lacking. Proper demand management facilitates the planning and use of resources for profitable business results. It encompasses the activities of forecasting, order entry, order promising, and determining branch warehouse requirements, interplant orders, and service parts requirements.

APICS Dictionary, 9th edition, 1998

Production plan by month (in dozens)

	January	February		March	....
Mens Shirts	200		200	150	

	Week 1	Week 2	Week 3	Week 4
Striped, long sleeve, 17 - 171/2, RS329	20	20		
Striped, long sleeve, 16 - 161/2, RS329	20	20		
Striped, long sleeve, 15 - 151/2, RS329	20	20		
Striped, long sleeve, 14 - 141/2, RS329			10	10
Solid, long sleeve, 17 - 171/2, RS500			20	20
Solid, long sleeve, 16 - 161/2, RS500			10	10

Figure 9.3   MPS Is Constrained by Production Plan

The top level hierarchical plan becomes the input for the next level of planning, which is the master production schedule (MPS) or master schedule level. This schedule is always prepared in terms of end items (i.e., what is shipped from the factory or plant). If the factory produces bicycles, then the end item is a bicycle. But, if the factory produces front wheels for another factory that assembles them onto frames, then the end item is the front wheel. Planning for the MPS is constrained by the earlier planning which resulted in the production plan. Remember that the production plan was developed in terms of product families. The MPS is very specific about the end item. For example, the production plan for an apparel factory may have scheduled 200 dozen men's long sleeve shirts for the month of January. The production plan did not specify color or design, just the family of shirts that would be produced. The MPS might then divide up these 200 dozen into specific end items to be produced in specific time periods. This is illustrated in Figure 9.3 for the month of January. Note that the shirts that are scheduled to be produced in all of the weeks of January total the 200 dozen scheduled to be produced in January.

Master production schedule (MPS)—1) The anticipated build schedule for those items assigned to the master scheduler. The master scheduler maintains this schedule, and in turn, it becomes a set of planning numbers that drives material requirements planning. It represents what the company plans to produce expressed in specific configurations, quantities, and dates. The master production schedule is not a sales forecast that represents a statement of demand. The master production schedule must take into account the forecast, the production plan, and other important considerations such as backlog, availability of material, availability of capacity, and management policies and goals.

APICS Dictionary, 9th edition, 1998

If the factory being scheduled is an assembly factory, then the MPS is the Final Assembly Schedule (FAS). This creates the sequence in which work will come down the production line.

Final assembly schedule (FAS)—A schedule of end items to finish the product for specific customers' orders in a make-to-order or assemble-to-order environment. It is also referred to as the finishing schedule because it may involve operations other than just the final assembly; also, it may not involve assembly, but simply final mixing, cutting, packaging, etc. The FAS is prepared after receipt of a customer order as constrained by the availability of material and capacity, and it schedules the operations required to complete the product from the level where it is stocked (or master scheduled) to the end-item level.

APICS Dictionary, 9th edition, 1998

Given that the planner has developed an MPS, the planner needs to check whether this MPS is feasible (i.e., is there enough capacity to actually do the work that the MPS is scheduling the plant to do?). It is crucial that the MPS be feasible, because the remaining planning steps all assume that the MPS is reliable. If the MPS is not feasible, then none of the following plans is valid. The feasibility of the MPS is checked via rough-cut capacity planning (RCCP). The RCCP process is for the medium term, which depending on the company could be 1 week or 3 months. If there is not enough capacity to produce what is on the MPS, then some work scheduled on the MPS has to be pushed ahead to where there is excess capacity or the work has to be postponed.

Rough-cut capacity planning (RCCP)—The process of converting the master production schedule into requirements for key resources, often including labor, machinery, warehouse space, suppliers' capabilities, and, in some cases, money. Comparison to available or demonstrated capacity is usually done for each key resource. This comparison assists the master scheduler in establishing a feasible master production schedule. Three approaches to performing RCCP are bill of labor (resources, capacity) approach, the capacity planning using overall factors approach, and the resource profile approach.

APICS Dictionary, 9th edition, 1998

Master Production Schedule

It is critical that firms be able to plan and schedule production in their own firms accurately for supply chain integration to work. The master production schedule is the outcome of all of the planning. It is the contract between one firm and its marketing department and hence the remainder of the supply chain. If one member of the supply chain cannot create a valid master production scheule and adhere to the schedule, then there is a large amount of variance within the chain and all remaining members of the supply chain must include a large amount of either spare capacity or lead time to ensure that they deliver when they agreed to.

As supply chain management develops, companies will link their demand and supply functions through integrated forecasting, distribution planning, and the master planning process. This collaborative planning is only possible if each member of the supply chain performs its portion of it correctly. In this environment it is necessary for the schedule to react dynamically to demand and supply changes.

Once a firm has a feasible MPS, the managers proceed to the next phase of planning, which is materials requirements planning. To produce the MPS end items a firm needs to procure or produce intermediate materials. For example, a bicycle wheel is assembled from a hub, spokes, and a rim. The material needed to produce the items in the MPS is calculated at the Materials Requirements Planning stage in Figure 9.2. The different methods to do this will be considered later. Information from the detailed material plan is used to determine if there is adequate capacity in the detailed capacity planning activity. In this stage the load on each resource in each time period is determined and the schedule may be revised or capacity adjusted if necessary to ensure that the schedule is feasible. The detailed capacity plan and the detailed material plan are used to create a set of material and capacity plans that are feasible.

The material requirements plan becomes input to the shop-floor system, which is the unit that actually starts to manufacture the products. The detailed plans are also sent to the vendors to ensure that they provide the needed materials in time for production of the end product.

In Figure 9.2, the shop floor control function is shown by the box labeled "Production Activity Control." This is a short-range tracking function, which watches the progress of jobs through the shop, determines when to release more work (Input/Output Control), and determines the specific sequences of conducting work at the individual work centers.

Influence of Supply Chain on Demand Management

The relationship the members of the supply chain have with each other is strongly represented in how the firms perform the demand management function. In supply chains where information is shared quickly and the information is accurate, the demand management function of those firms who do not serve the end customer is less one of forecasting and more one of responding to known needs of its customer, the next firm in the supply chain. When the relationships in the supply chain are more competitive or at arm's length, firms must actively manage demand. For example, in the appliance repair industry, individual firms will obtain the parts they need for repairs from a large number of distributors or if possible directly from the factory. The distributors will try to influence demand by offering different pricing arrangements, a variety of discounts, and possibly quantity discounts in an attempt to obtain the largest share of a business. This type of active market manipulation requires close coordination between marketing and operations to ensure that as demand is stimulated it can be met.

Demand management includes forecasting demand, accepting orders, and processing orders. It also includes stimulating demand when it is low.

Supply Chain and Resource Planning

Resource planning is concerned with providing the necessary amount and type of capacity needed to serve the demand in the market. Many of the resource planning decisions are long range. These include plant and equipment decisions. In the near term, resource planning is concerned with labor availability. Resource plans serve as constraints on the production plan.

The resource planning conducted by a firm is influenced by the type of supply chain that the firm is involved with and the firm's own business strategy. For example, if a firm is a member of a supply chain where the number of suppliers of its products is being reduced and the firm was selected as one of the partners to receive more volume, the firm would have an incentive to invest in more capacity. If, on the other hand, the firm is one of many generic suppliers of a product in a supply chain where the customer treats all competitors' products as commodities, the firm would have an incentive to compete on the basis of price. This would cause the firm to resist investment in capacity with a long pay back period and instead to focus on reducing other costs.

Supply Chain and Production Planning

Production planning coordinates manufacturing/operations with the firm's business plan. It develops an aggregate plan that may be stated in dollars or aggregate units of products. For example, it may be stated in one firm in terms of cars to be sold and in another in terms of tons of fabric to be sold. This is not a detailed plan. All the products of the firm are often grouped into families. This is the planned production of the family groups that will meet the expected demand.

One advantage of a tight, highly communicative supply chain is that the production plan is based less on a forecast and more on actual demand. In tightly communicative supply chains firms may share their production plans with each other to encourage coordinated planning into the future between all of the members of the supply chain. In this type of environment it is much more likely that all sources of demand have been accounted for and there will be no surprises that upset the production plan.

Firms frequently update the production plan to recognize changed conditions in the market. They may do this by reviewing the plan monthly and updating it quarterly. Typically the production plan is prepared for the next 12 months.

Supply Chain and Master Production Schedule

The master production schedule (MPS) is the detailed production plan for the end items to be produced by the plant or firm. It is constrained by the production plan. For example, if the production plan calls for 1,000 two-liter size bottles of soda to be produced in January, then the MPS is the detailed plan to produce all the different types of two-liter size bottles of soda produced in January. This may mean that 250 two-liter bottles of cherry cola are produced in week 1; 250 bottles of diet cola in week 2; 250 bottles of sugar cola in week

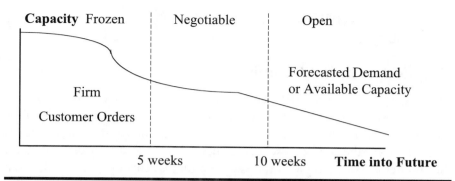

Figure 9.4 Time Fences in Master Production Schedules

3; and 250 bottles of cherry cola in week 4. This process of identifying the amount of each specific end item to be produced in each time period is called *disaggregation.*

The MPS is changed and updated more frequently than the production plan. It is a detailed plan and does not extend as far into the horizon as the production plan. It is often viewed as a contract between production and the rest of the firm. In the time horizon in which the MPS is frozen, production will produce everything that is stated on the schedule and the rest of the firm will sell and/or accept what is produced. Firms set up policies about when and the extent to which an MPS can be revised. One example is shown in Figure 9.4. Here the MPS is broken into 3 areas. One where it is frozen and no changes are allowed, one where changes are negotiable, and the third where changes can be made without penalty. This agreement to have time fences is to achieve coordination both within and outside the firm. It creates a clear understanding of the flexibility of the firm to respond to changes. The time fences should represent areas where actual constraints begin. For example, the frozen section of the MPS should represent an area where production has already started, or the firm is committed to the receipt of the raw materials ordered from the supplier. The determination of where these time fences are to be placed is a strategic issue, since it will influence the relationship of the firm to the other members of the supply chain. These decisions determine the overall ability of the supply chain to respond to changes in customer demand.

For example, if a textile company supplies fabric to a manufacturer of office furniture and the textile manufacturer has a 12-week time fence, the flexibility of the office furniture manufacturer is limited. It is difficult a month before delivery to respond to changes from the customer who may want to change the colors of the seats on the desk chairs previously ordered. However, if the fabric manufacturer has a 3-week time fence, then color changes can occur close to the time of delivery, which allows the end customer, who may be decorating the entire office, to select a more suitable color.

Detailed Material Planning

The essential point about detailed material planning is that there is no forecast involved. All demand for material at this stage of the planning hierarchy is derived from the demand for the end items. The demand for the material components of the end items scheduled in the MPS is said to be dependent on the end items. For example, if the end item that is scheduled in the MPS is a bicycle, then we know that we need two wheels and one seat, etc. We can derive the demand for the wheels that we need from the information that we have about the need for end items.

> **Dependent demand**—Demand that is directly related to or derived from the bill of material structure for other items or end products. Such demands are therefore calculated and need not and should not be forecast. A given inventory item may have both dependent and independent demand at any given time. For example, a part may simultaneously be the component of an assembly and sold as a service part.
>
> *APICS Dictionary*, 9th edition, 1998

A bill of materials (BOM) is illustrated in Figure 9.5. The highest item in the BOM is the end item that is scheduled in the MPS. This end item is always placed at the top of the BOM in what is designated as level 0. The BOM is basically the list of ingredients to manufacture the end item. It is read in descending order beginning with level 0. To make 1 A, we need 2 Bs and 1 C (notice the numbers given in brackets next to the letter inside the circle). As we read down the BOM we often use the terminology of parent–child. In the BOM an item which is composed of one or more components is referred to as the parent of those components. The components are, of course, referred to as the children. In this BOM both B and C are children of A. B is itself a parent of D and E who are its children. To make 1 B, we need to have 1 D and 2 Es. Finally, to make each E we need to have 3 Fs. C, D, and F do not have any children in this BOM, because they are purchased outside of the firm. That does not mean that they are simple parts. For example, C could be a master cylinder for a brake assembly that we purchase from someone else.

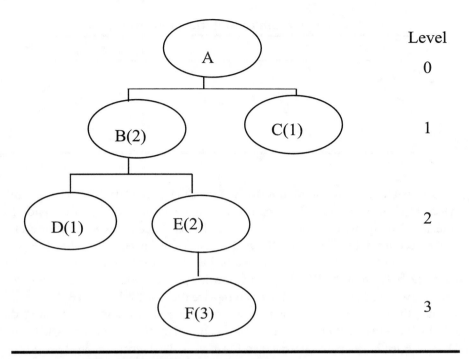

Figure 9.5   Bill of Materials (BOM) for Item A

Another form of the BOM is called the indented bill of materials. It is illustrated in Table 9.1. Notice that the higher levels of the BOM shown in Figure 9.5 are in the columns farthest to the left of Table 9.1. The indented BOM provides the same information as the BOM presented earlier, but in a different format. The quantity of each child needed to produce one parent is shown in the column the farthest to the right.

Indented bill of material—A form of multilevel bill of material. It exhibits the highest level parents closest to the left margin, and all the components going into these parents are shown indented toward the right. All subsequent levels of components are indented farther to the right. If a component is used in more than one parent within a given product structure, it will appear more than once, under every subassembly in which it is used.

APICS Dictionary, 9th edition, 1998

Table 9.1 Indented Bill of Materials for Product A

Level 1	Level 2	Level 3	Number Needed
B			2
	D		1
	E		2
		F	3
C			1

To ensure you understand how to read the BOM, calculate how many Fs are needed to produce 5 As if there are no existing inventories of either the end or intermediate items. To produce 1 A we need 2 Bs. Since we need 2 Es for each B, we need a total of 4 Es (i.e., 2Es × 2 Bs). Since we need 3 Fs for each E we need a total of 12 Fs (i.e., 4 Es × 3 Fs). If this is confusing, it is useful to work from the bottom of the BOM upwards for this particular example. In this case because we need 3 Fs for each E we can replace the E with the term 3 Fs so that we have the term 3 Fs(2) as shown in Figure 9.5. This term can then be changed to 6 Fs. Working further up the BOM, we can replace the term B with the term D(2) + 6 Fs(2) since 2 Bs are needed for each A. This term can be changed to D(2) + 12 Fs, demonstrating that we need 12 Fs to produce each A.

Detailed Capacity Planning

While the BOM contains a list of the ingredients needed to produce an item, it does not tell when or how to combine them. To determine the capacity that is needed to manufacture an end item in the MPS, we need to know what types of capacity are needed and how much capacity is needed. This is contained in the routing sheet. The routing sheet (also referred to as the routing or the route sheet) contains the basic information for the flow of the product through the shop.

Routing—Information detailing the method of manufacture of a particular item. It includes the operations to be performed, their sequence, the various work centers involved, and the standards for setup and run. In some companies, the routing also includes information on tooling, operator skill levels, inspection operations, and testing requirements, etc.

APICS Dictionary, 9th edition, 1998

Operation sequence and processing data

Operation		Standard Times (hours)		
Sequence No.	Description	Processing	Setup	Component
1	Shear	0.10	0.40	C
2	Blank	0.08	0.50	D
3	Inspect	0.06		E
4	Saw	0.33	1.40	F
5	Heat Treat		10.00	G
6	Subassemble C&D	1.3		B
7	Sand Blast	0.28	0.25	G

Figure 9.6   Routing Sheet—Operation Sequence and Processing Data

An example of a routing sheet is shown in Figure 9.6. In this example the setup times and processing times for an individual part are given. The routing sheet also gives the sequence number for the product through the shop.

A routing sheet for the BOM example above is given below in Table 9.2. The path that the part is to follow is given in the left hand column. In this example there are five operations needed. The first and second operations occur at work center 100. The estimated time needed for both set-up and processing is given. The first step is to unpack the component and the second is to clean and inspect it. The third operation is to put 3 of the parts into a jig and to weld them together. The fourth operation is to inspect the component again. The fifth operation is to drill 3 holes.

Table 9.2   Sample Routing Sheet

Operation	Work Center	Description	Setup Time	Processing Time Per Unit
1	100 Inspect	Unpack	0	30 sec.
2	100 Inspect	Clean and inspect	0	45 sec.
3	200 Weld	Place 3 Fs into jig and weld	15 min.	180 sec.
4	100 Inspect	Inspect	0	30 sec.
5	300 Drill	Drill 3 holes	30 min.	120 sec.

Given the route, which tells to which work centers the part will go and the amount of resources needed at each resource, plus the material plan the operations planner can determine how much capacity is needed. This process is shown in Figure 9.2 as Capacity Requirements Planning. At this stage the planner calculates the total amount of each type of resource needed in each time period. This step is done to ensure that the plant is creating a feasible plan (i.e., one that can be successfully implemented). Often software is used to do the CRP step, but for small shops it can be done by hand. Given the calculations above, the planner now knows how much work is required from each work center in each time period. These loads are summed by work center by time period. To demonstrate, this is illustrated with the load created by the part described in Table 9.2, which has an order size of 100 parts in Period 1. The load for Period 1 at each work center is given in Table 9.3. Work center 100 has 105 seconds of work for each of the 100 parts to be produced. This is a total load of 175 minutes. Work center 200 has a load of 315 minutes (i.e., 15 minutes + 100*180 seconds/60 seconds/minute). Work center 300 has a load of 230 minutes.

Table 9.3   Work Center Load for 100 Parts for 1 Period

Work Center	Load (Minutes)
100 Inspect	175
200 Weld	315
300 Drill	230

The calculations in Table 9.3 show that the shop load is very low, so the planner can add more work. At the completion of this task the planner has a detailed set of material and capacity needs with their due dates. The next stage is to actually execute the plans.

The planner now releases the job to the shop floor, and orders for purchased materials are released to the vendors. The shop floor control system controls the actual flow of the work through the shop floor.

Capacity planning is essential to successfully fulfill the plan. If work centers are overloaded, then they cannot produce all of the product that they are expected to produce. When one work center fails to meet its goals, all of the work centers fall behind and the plan is not met. To be able to successfully meet the goals of the plan, the plan must be feasible. However, it is not always necessary to do detailed planning on all of the work centers to determine if the plan is feasible. This is particularly true when shop managers know that several of the work centers tend to be the bottleneck. These critical work centers can then be used to evaluate the feasibility of the plan. This is done by checking the load at each of these work centers. If these work centers are not overloaded, then it is likely that the plan is feasible. This may not be true if the bottlenecks shift around the plant a lot. The bottleneck may shift due to product mix changes, which may be themselves due to seasonal fluctuations in demand for different products. While this rough-cut capacity planning has its limitations, the detailed capacity planning which examines each work center is also constrained. The output of a detailed capacity planning calculation depends on the value of the input. In a large shop, it may be very difficult to obtain accurate data on each work center. This limits the value of the detailed capacity requirements plan.

In the prior example, the rough-cut capacity planning would look at just work center 200–welding, because it is the most constrained. Rough cut considers only the total load on a work center during a specified time period such as a week. So, this one job would add 325 minutes of work to work center 200 in week 1. If the other jobs contributed 2,000 minutes of work in this time period, work center 200 would have 2,325 minutes of work. At 1 shift/day, there are 420 minutes/day of capacity, or 2,100 minutes in a 5-day work week. So, this analysis says that work center 200 is overloaded and some work needs to be shifted somewhere else.

Just-in-Time (JIT)

Just-in-Time (JIT) production is a philosophy rather than a clearly defined method to control production. What most managers consider to be JIT, originated at Toyota where it is referred to as the Toyota Production System. This system is still evolving as computer integrated manufacturing technology and information systems are integrated with it. The details of Toyota's system are provided by Monden in his book the *Toyota Production System: An Integrated Approach to Just-in-Time* (1993). From 1973 when the Toyota system began to be widely adopted in Japan (due to the oil shock) until 1983, JIT received a limited amount of attention in the United States. With the publication of Robert Hall's book *Zero Inventories* (1983), the concepts and practices of JIT received wide acceptance and adoption by many firms. JIT is defined as a philosophy of waste reduction, not a limited set of practices. Schneiderjans (1993) has examined all of the literature about JIT and has determined that there are eight key principles. These are given in Table 9.4.

Table 9.4 JIT Principles

1. Seek a produce-to-order production schedule.
2. Seek unitary production.
3. Seek to eliminate waste.
4. Seek continuous product flow improvement.
5. Seek product quality perfection.
6. Respect people.
7. Seek to eliminate contingencies.
8. Maintain long-term emphasis.

Adapted from Schneiderjans, 1993.

At first glance, most of these principles may seem to oppose JIT's philosophy of waste elimination. But, they actually do promote waste elimination. By seeking a produce-to-order system, a firm is able to eliminate the waste of finished goods inventory, because the product is sold when completed. By seeking unitary production a firm eliminates the waste of both work-in-process and finished goods inventory. A side effect of this is that a firm using unitary production is able to respond to customer changes much more quickly, since the lead times are shorter. Because there is no buffer of inventory in the system, unitary production reveals problems that might be hidden by the inventory. The third principle is to eliminate waste by using only the minimum amount of equipment, materials, and human resources required. Seeking continuous product flow improvement eliminates bottlenecks and tries to balance all of the processes, thus eliminating idle time. Under this principle all activities that are not required for work are eliminated. By seeking perfect quality JIT eliminates waste by building 100% perfect products. Respect for people involves workers in the production control, giving them the authority to insist on perfect quality. By eliminating contingencies, managers reduce inventory, which reduces waste. A long-term emphasis is required since JIT's benefits can take a long period of time to mature.

It is important to understand that JIT is not an inventory reduction system. It is a system that focuses on the elimination of waste in all of its forms. It is sometimes thought of as stockless production, because it uses much less inventory than other systems. But, it still requires inventory. The level of inventory does provide a good measurement of how effectively waste is being eliminated. In this way, inventory is analogous to a thermometer. A thermometer tells the temperature, but it does not create the temperature. The level of inventory in a system tells the observer how effectively waste has been

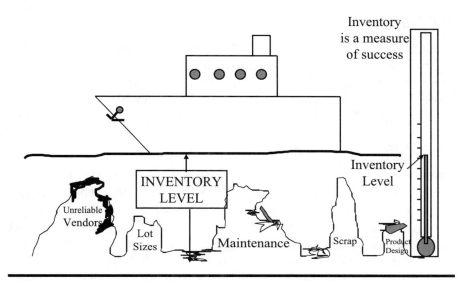

Figure 9.7   Inventory as a Measurement of Success

eliminated. The reduction of waste did not eliminate the waste. This is illustrated in Figure 9.7. In this figure it is also illustrated that inventory is often used in a system to hide problems or to hide waste. Inventory is like the level of water in a stream, which allows a boat to float safely over the rocks and other obstacles. If we remove the rocks the boat can safely float in a lot less water.

It was only possible to operate with less inventory after the problems creating the waste in the system were eliminated. Inventory is an effective measure of waste in the system, because the level of inventory in the system is the combined outcome of all the system components. If the quality of the system is poor, then there is additional inventory to compensate for the bad quality, so bad parts will not shut the system down. If the process used in the system is defective, then additional inventory is needed to accommodate the problems such as an extended flow time. The product design greatly influences inventory. The design of a product determines how many components there are in a product and also determines how easy it is to manufacture the product. Maintenance certainly influences the level of inventory, as machines, which are not maintained appropriately, break down more frequently and consequently more inventory is needed to protect against breakdowns. The employees' motivation makes a difference in the level of inventory in a shop as the employees can help solve problems if they are motivated to do so.

JIT Tools

There is a lot of confusion between JIT the philosophy and the tools used by JIT to achieve its goal of waste elimination. There is a distinction between these two. Any firm can use some of the JIT tools and not create a JIT system. These tools are techniques to fulfill the principles defined in Table 9.4. One of the most widely known JIT tools is the pull system.

Pull System

Often when people talk about JIT they refer to it as a *pull system.* By this they mean that material is pulled through the production system. The illustration in Figure 9.8 shows that inventory is replenished throughout the system only in response to customer demand. There is component inventory before each work center and only when that component inventory decreases to a certain point does the next work station start to replace the inventory.

There are many variations of the pull system. But, they all work by basically setting a limit on the number of parts to be placed into a container (e.g., a kanban container) and limiting the number of containers that there can be between work stations. One of the simplest pull systems is to draw a square on the floor or table between two work stations and limit the amount of work that can be in that square to 1 item. The preceding work station cannot produce until the square is empty. It then produces enough to fill the square and then stops. This ensures that all the work stations in a line work at the same pace. This is illustrated in Figure 9.8 where work station A does not work when there is a part in box AB. When work station B needs work, it pulls one item out of box AB. At this point work station A will begin work.

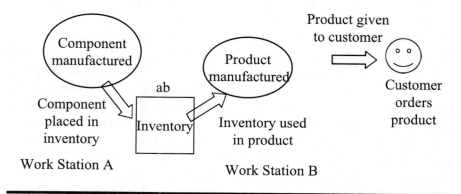

Figure 9.8   JIT Is a "Pull System"

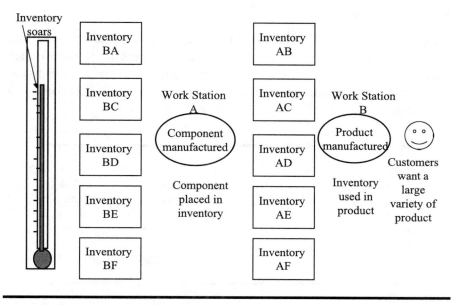

Figure 9.9   Limits of Pull System

If JIT is a philosophy, then the pull system is one JIT tool. It is a method of eliminating waste in the movement of material. As a tool, it is important that managers understand when and where it is useful and where it will create problems. The pull system simplifies the flow of material through the shop. Material moves from point to point only when needed. It also moves only in the quantities needed. It is a self-limiting system. When a shop produces standard products that ship as discrete units, it is feasible to use the pull system. However, if a shop produces a wide variety of products, of limited quantity and highly variable demand, then the pull system may increase the inventory of the shop. This is illustrated in Figure 9.9 where the customer wants a wide variety of product. To produce these quickly the shop maintains a wide variety of components. But, to maintain the wide variety of components, a wide variety of materials must be kept before work station A to produce the components. Much of this inventory may sit for long periods because there will be no demand for it from the customers. If the production chain is long, a large amount of inventory is necessary to support this pull system. All of this inventory costs money and takes up storage space. So, one environmental requirement for JIT is that there be relatively smooth demand for the end item.

Setup Reduction

Setups cost money and take time. If the setup takes too much money, then there will not be enough capacity at a work center to produce all the products required. When setups take too long at a work center, it is not possible to implement the pull system without reducing the setup time. This concept is sometimes called SMED or single-minute exchange of die.

Setup—1) The work required to change a specific machine, resource, work center, or line from making the last good piece of unit A to the first good piece of unit B. 2) The refitting of equipment to neutralize the effects of the last lot produced (e.g., teardown of the just-completed production and preparation of the equipment for production of the next scheduled item).

APICS Dictionary, 9th edition, 1998

A common approach in setup reduction is to identify each step in the entire setup process and all the tools and skills required to do this. The process may be videotaped or a flow chart may be created. This is illustrated in Figure 9.10.

Part Standardization

Another tool that is vital to implement the pull system is the standardization of parts. As we saw earlier, too many parts increases inventory throughout the

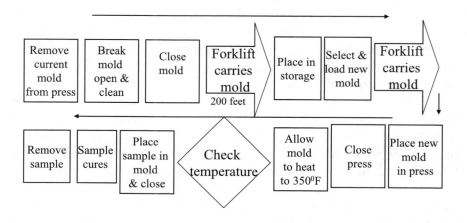

Figure 9.10   Example of Setup Process at Rubber Press

system. If parts can be standardized across parts, then less inventory needs to be carried. By standardizing parts, one part can be used in multiple models or components. This means that one store of parts can serve multiple products. This allows a reduction in inventory, because the amount of inventory needed is a function of the square root of demand. If demand doubles, then inventory needed increases only by the square root of 2.

Another tool is to standardize processes. This has multiple benefits. One is that repetition increases learning and decreases defects. Another is that setup times can be reduced. Both of these allow inventory to be reduced.

Creating linear product flows helps to standardize the process. A linear product flow as described earlier means that product can move in single units in a linear path. The product does not have to loop back to a prior work station. One tool in creating a product flow is cellular manufacturing. To achieve cellular manufacturing a firm must group or organize their parts by similarities. The process of doing this is called group technology (GT).

Usually parts have grouped characteristics that determine the type of processes they need. Those parts that require similar processes are grouped together and assigned to a cell.

Cellular manufacturing—A manufacturing process that produces families of parts within a single line or cell of machines operated by machinists who work only within the line or cell.
APICS Dictionary, 9th edition, 1998

Group technology—An engineering and manufacturing philosophy that identifies the physical similarity of parts (common routing) and establishes their effective production. It provides for rapid retrieval of existing designs and facilitates a cellular layout.
APICS Dictionary, 9th edition, 1998

By assigning a group of parts to a cell, there is sufficient volume to create a product flow for the set of parts. In addition, the cells can be arranged in a loose product flow. So, there may be one cell, which prepares raw materials for use in 5 or 6 other cells. Those cells may take that input and finish the process including the boxing and shipping. This is illustrated in Figure 9.11. In this plant all the products require common raw materials which need to be prepared just before their use. Each of the raw materials prepared may be different, but they are similar enough to do in the raw material preparation cell. The cells produce families of similar products, package them, and then take them

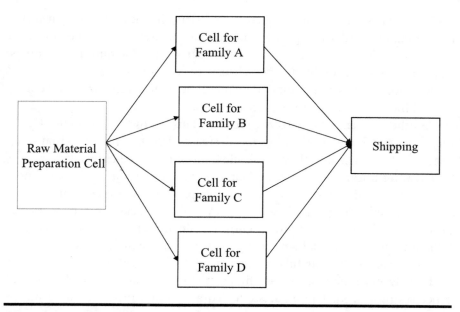

Figure 9.11   Cellular Layout Example

to the shipping dock. All the material flows are linear without looping, even though there is a large variety of product being produced.

Uniform Master Production Schedule

Another tool of JIT is the uniform master production schedule. The uniform MPS can be useful in any type of production planning and control system, but it is only feasible when there is a set of standard products with relatively high demand. The uniform MPS means that there is little variation in the production quantities between days. This requires the planning department to level the load. The advantage of being able to create this uniform MPS is that it drives the entire system. So, the rest of the manufacturing system will have much less variance in it because of the load leveling at the MPS level. This should increase productivity because bottlenecks will not float around the shop disrupting material flow.

Given the uniform MPS, the JIT planner divides the monthly production into daily production schedules. It is important that each day have the same production level. Further it is important that the production runs approach single unit batches if feasible. This leads to a very smooth flow of material.

To achieve this daily uniform schedule, the schedule must do mixed model scheduling when the facility is producing more than one product. The goal in the sequence of production is that two identical products are not produced next to each other. For example, in a JIT shop, the capacity is 420 minutes/day and there are 20 days in the month. Three models are being produced (A, B, and C). The demand for A is 4,800, the demand for B is 2,400, and the demand for C is 1,200. To develop a level schedule for these three models, first translate the monthly demand into daily demand as shown in Table 9.5. Second, calculate the cycle time necessary to produce all the models. Remember that cycle time (c) is the time to complete one unit, or the time between the completion of each model. In this case 420 units need to be produced in 420 minutes, so c = 1 minute/unit. With 420 minutes in a day, for Model A, c = 420/240 = 1.75. This means that every 1.75 units on average the plant needs to produce a model A.

Table 9.5   Mixed Model Scheduling Example

Model	Demand/Month	Daily Demand	Average Time Between Models (c)
A	4800	240	1.75
B	2400	120	3.0
C	1200	60	7.0

The next step is to find a sequence which allows all of these products to be produced yet does not require a product to be produced in sequence next to another of its own products. To set up the sequence, one heuristic is to schedule the product with the longest average time between models first. Create a time line as shown in Table 9.6 where the time measure is based on the overall cycle time needed. Schedule C first with 7 minutes between production. This means that Model C is produced at minute 1 and minute 8. Then schedule Model B every 3 minutes. So, if B starts at 2, another is scheduled at 5 and another at 9 since there is already a C being produced at minute 8. But, this would leave an A being produced at minutes 3 and 4 then 6 and 7. To split production of the A models more, B could start in minute 3, then minute 6. So, the final schedule, a portion of which is shown in Table 9.6, repeats the sequence of C, A, B, A, A, B, A. It is clear that a C is produced every 7 minutes. Since the cycle repeats every 7 minutes, it will be done 60 times a day. So, 240 A are produced (60*4), 120 B are produced (60*2), and 60 C are produced (60*1).

Table 9.6   Time Line

1	2	3	4	5	6	7	8	9	10	11	12
C	A	B	A	A	B	A	C	A	B	A	A

Worker Involvement

A high degree of worker involvement is important in a JIT facility. The workers initiate many actions such as turning on *andon* lights when there are problems. These lights are used by workers to signal problems.

Andon—1) An electronic board that provides visibility of floor status and provides information to help coordinate the efforts to linked work centers. Signal lights are green (running), red (stop), and yellow (needs attention). 2) A visual signaling system.

APICS Dictionary, 9th edition, 1998

Another example of the importance of the work force to JIT is the need to cross-train workers. Cross-training is when workers are trained to do more than one job. A cross-trained worker can help someone at the next workstation if that worker falls behind and a cross-trained worker understands what the quality requirements are of those processes that the worker is feeding. A cross-trained work force is flexible. It is possible for the workers to move to where they are needed and to respond on short notice. Job rotation is one method companies use to ensure that workers retain their training skills. A firm may set up a job rotation system where an employee routinely rotates between 3 different jobs to ensure that his/her skills remain sharp.

A common technique to achieve worker involvement is to set up work teams. Techniques to do this are discussed in later chapters.

Cross-training—The providing of training or experience in several different areas, e.g., training an employee on several machines rather than one. Cross-training provides backup workers in case the primary operator is unavailable.

APICS Dictionary, 9th edition, 1998

Job rotation—The practice of an employee periodically changing job responsibilities to provide a broader perspective and a view of the organization as a total system, to enhance motivation, and to provide cross-training.

APICS Dictionary, 9th edition, 1998

MANAGING FOR IMPROVEMENT IN THE SUPPLY CHAIN

IV

10 Planning, Controlling, and Improving the Supply Chain

Total Productive Maintenance

Total Productive Maintenance (TPM) has evolved over the past 30+ years to ensure equipment reliability at low cost. TPM emphasizes zero breakdowns and zero defects. It supports the philosophy of JIT to eliminate waste. TPM uses all the tools that have evolved over the past 100 years to improve equipment reliability, but it organizes the maintenance process so that it becomes a continuous learning process for everyone involved in it. For example, TPM is a set of techniques to manage breakdown maintenance, preventive maintenance, and predictive maintenance. It applies the scientific method to the management of these programs and to the management of equipment. The goal of TPM is to create an environment in which people and equipment can deliver exactly what the customer requires without waste. TPM actually creates a new system for equipment management. It depends on autonomous teams and on cross-functional teams for its employment.

> **Total productive maintenance (TPM)**—Preventive maintenance plus continuing efforts to adapt, modify, and refine equipment to increase flexibility, reduce material handling, and promote continuous flows. It is operator-oriented maintenance with the involvement of all qualified employees in all maintenance activities.
> *APICS Dictionary,* 9th edition, 1998

The other tools that are used with JIT are often used with many other systems and are discussed separately. Some of these include total quality management, supplier certification/management, and supply chain management.

Continuous Improvement

JIT cannot be successful without continuous improvement. As part of the effort of achieving continuous improvement, it must include total quality control. To achieve this it must have employee involvement. One method used by firms to achieve or to help highlight their progress towards these goals is inventory reduction. This can quickly make any issues in production evident. There will not be any inventory to hide the problems.

The first step to achieve continuous improvement is to develop a process view of the system. This means being able to understand the interaction of all the separate parts. The second step is to understand that individuals respond to the performance measures that are put into place, so the firm needs to ensure that there are performance measures, which encourage continuous improvement. This is illustrated in Figure 10.1 where process or improvement criteria are in place for each step in the process. Results measures are also in place to measure how well the process produces an outcome. But, the results measures are not the sole performance of the system's success. The system is successful only if it is producing good results and improving. For example, if the process is an apparel manufacturer who is trying to reduce the order cycle time, process measures may be flow time measures or queue time measures. The results measures may be the percent shipped on time. There is a logical connection between the process measures and the hoped for result of shipping on time.

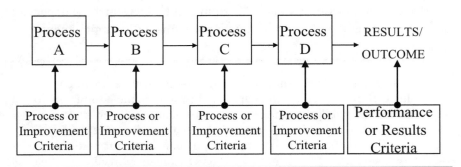

Figure 10.1   Process View of System (Adapted from Imai, 1986, p. 18)

Focused Factory—JIT Tool

A focused factory limits its production to products whose characteristics match the firm's competitive strategy and existing technology. This means that the factory has specialized to produce a narrow range of products. For example, it might produce gear reduction boxes for conveyor systems from purchased castings and steel. But, it will not produce any other portions of the conveyor system, because these other portions would require different skills and different technologies.

Focused factory—A plant established to focus the entire manufacturing system on a limited, concise, and manageable set of products, technologies, volumes, and markets precisely defined by the company's competitive strategy, its technology, and economics.
APICS Dictionary, 8th edition, 1995

The reason that factories are focused on a narrow product area is so they can do a better job of meeting market needs than a factory producing a wide range of products. The focused factory is able to do this in two ways. First, the factory is designed so that there is a lot of repetition allowing both the workforce and the managers to learn the best methods of producing the product. Second, because there is a limited number of products, the materials inventory required to produce the products is reduced. This allows the firm to reduce the queue lengths so that the production lead times are less.

A drawback of the focused factory is that if market conditions change so that there is not sufficient volume needed of the product it had been producing, then it may no longer be cost effective. One approach to create flexibility while gaining the efficiency of specialization is to create a "plant within a plant." In this setup managers treat different work areas within a plant as if they were different factories, even though they are sharing the same infrastructure. For example, a factory that produces molded rubber parts could divide itself into several parts. Factory 1 could produce the mixes of rubber needed by the other factories. Factory 2 could produce molded rubber parts for the auto industry, and Factory 3 could produce molded rubber parts for lawn mowers. Each factory could have a separate manager, all reporting to a site manager. But, each factory would share the same infrastructure. There would be one MIS system, one accounting office, one switchboard, and one maintenance staff. Each factory would receive its own income statements and balance sheet. The advantage is that each factory focuses on its own customers (internal or external) to obtain efficiencies due to this specialization. However, if there is a downturn in one factory, the excess workers can be hired by another factory. So, there is increased flexibility as the work force can shift to the areas with the greatest demand.

Simplification

Simplification is an important part of improving a system. Simplifying a system by reducing the number of activities, reducing the complexity of activities, or improving the layout increases the rate of learning and decreases the potential for errors. Often systems are complex because there are unresolved problems. It is a paradox, but simple systems require managers to have more insight into the process than complex systems. Many of the tools examined elsewhere in this book are used to simplify the system. Some of these tools are the cause-and-effect diagrams, layout improvement, and process flow diagrams.

Standardization

There are several advantages of standardization. One is that parts are interchangeable. The parts for one model of machine will fit a second machine of the same model. Another advantage is that communication with the customers is simplified. When the customers know that a standard range of sizes, features, colors, or other attributes is available, they can select one and not have to engineer another set of specifications.

Standardization—1) The process of designing and altering products, parts, processes, and procedures to establish and use standard specifications for them and their components. 2) Reduction of total number of parts and materials used and products, models, or grades produced. 3) The function of bringing a raw ingredient into standard (acceptable) range per the specification before introduction to the main process.

APICS Dictionary, 9th edition, 1998

Standardization has a large influence on the process. As the product is standardized so that one part may fit into multiple models, the total inventory needed by a firm is reduced. For example, if an automobile is assembled with 100 different types of screw fasteners, then the firm needs to stock all 100 types. If the number of fasteners is reduce by 50%, the firm can reduce its inventory by that same amount and focus on improving the quality of these fasteners. This allows the process to be simplified, because there are fewer items to be tracked and there are fewer installation methods, again leading to repetition and improvement.

11 The Synchronized Supply Chain

I n its simplest sense, the supply chain is nothing more that a holistic view of the business enterprise from the origin of raw materials through to the use of the completed product by the ultimate consumer. That process will likely involve multiple companies and may cross several industries. In the traditional methods of operation, each of those companies and industries will operate independently and in isolation from each other. Oftentimes, they will even treat each other as adversaries. Communication is limited to only the necessary transactional information required to create and fill orders from one company to and from the next link in the chain. Rarely is the information complete enough to offer any real opportunities for planning. There is a complete reliance on forecasting, which is notoriously inaccurate. In fact, the further back in the supply chain a company is, the less accurate the forecasting information will be. The root cause of all this is that the links of the traditional supply chain are considered independent rather than as parts of a single system.

Playing "The Beer Game"

An example of this phenomenon is made clear in an exercise called "The Beer Game" as described in *The Fifth Discipline* by Peter Senge. This is a board game using poker chips signifying inventory, preprinted playing cards indicating the fluctuations in demand at the retailer, and small forms on which the players write their orders to their suppliers. The scenario is that a retailer who sells beer is served by a distributor. The distributor is served by a warehouse, which is served by the manufacturer. Therefore, there are only four elements in this simple supply chain. There is a four-day lead time between each of the elements.

In the initial run of the game, each entity has a small stock of inventory, which is large enough to avoid any stock outs at the retailer assuming the forecast is correct and the lead times are met by each of the suppliers. Each entity is expected to receive orders from its customer and place orders for replenishment to its supplier. Other than receiving the orders, there is no communication allowed between the players. This is a simple, but realistic example of a supply chain.

As the game begins, the retailer "sells" four cases of beer each of the first six days. For the remainder of the 30 days of the simulation, the retailer demonstrates an increase in sales to eight cases of beer. As the distributor sees the inventory diminish after the first week of higher demand, he inevitably will increase his orders to the warehouse so that he can avoid any stock-outs. Accordingly, the warehouse sees these orders increase and decides to increase his demand on the manufacturer for the same reasons. The manufacturer, seeing an increase in demand, decides to work overtime to meet it. Because there is a four-day lead time, the increase in orders is not seen immediately, yet the succeeding step in the supply chain is threatened with missed due dates because of the increase in demand from its customer. Each of the elements becomes nervous seeing the inventory continue to decrease and backorders begin to climb. As a result, they further increase the order quantities to avoid missed deliveries and stock-outs. The effect on the manufacturer can be exponential as the order quantities are compounded at each preceding step. After a couple of weeks, the demand on the manufacturer can be hundreds of cases per day, yet the increase at the retailer was only four cases daily up to a total of eight cases.

Once this production finally begins to be filled from the manufacturer, the opposite effect occurs. The retailer, now having much more inventory than is needed, decides to order zero quantity for several days. This eventually is transmitted back to the manufacturer who is forced to stop production and may even lay off his workers. The stoppage then flows back downstream to the retailer and the whole fiasco starts all over again. These "waves" of high and low demand occur despite the fact that the real change in orders from the consumer was a delta of only four cases per day.

This scenario occurs in the real world every day. In an effort to avoid poor customer service, elements of supply chains do exactly the same thing. Ironically, the result of these actions is even worse customer service. The actual fluctuation in monthly demand in the automotive industry, for example, is less than 10%. However, second and third tier suppliers are constantly adjusting capacities by as much as 50%. In the textile and apparel industries, the effect of seasonal fluctuations only compounds the same problems.

Inevitably, demand on the manufacturing elements is much more severe than the measured effect on the retailer. The problem exists in every industry. The high and low waves of demand continue to occur despite the real demand of the consumer being relatively flat.

There Is a Solution

The concept of *synchronizing* the links of an entire supply chain to the demands of the market is the answer. In The Beer Game, this is demonstrated in the second run by simply allowing the elements of the chain to communicate the actual demand from the retailer on a daily basis. By ordering only the amount consumed each day throughout the system, the panic is avoided and the waves of demand are averted. *The solution is to treat the entire supply chain as one entity driven by the actual market demand.* Beer is brewed in the quantities each day equal to the amount sold at the retailer. Rather than each element operating in isolation, the entire system is synchronized to the market demand. The complicated paperwork between elements is eliminated. Each entity simply processes the amount pulled by the market daily. Orders are consolidated from all the retailers, and production schedules are planned accordingly. A buffer of inventory kept at the distributor absorbs any variability in the system. This variability can take the form of various "Murphy" events such as transportation delays, equipment breakdowns, and even weather problems. The size of this buffer of inventory is determined by the degree of variation and the amount of protective capacity within the system to absorb the variability. The greater the ability to respond with protective capacity, the lower the reliance on inventory. Such a system is able to operate in a much more stable manner and the stock-outs are eliminated (see Figure 11.1).

The Same Solution Will Apply to Any Supply Chain

This model is based on the concepts of *constraints management* and *synchronous flow,* which are rooted in the fundamental laws of physics. (See Chapter 7 for a further description of Constraints Management.) Focusing on a single control point and subordinating all other resources and processes to that point can synchronize any system. The obvious point to use as the control of a supply chain is the market that it serves. It makes no sense to produce or process any more than the market demands and it is fully intuitive that the entire system should be focused on producing just what the market wants. While this seems like a statement of the obvious, there is ample evidence that the business world does behave in this manner. To quote Mark Twain: "Common sense is not very common."

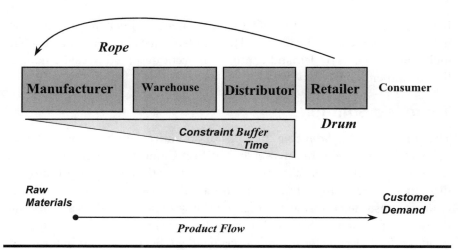

Figure 11.1   The Synchronized Supply Chain

The synchronized supply chain has been tested repeatedly and it does address the more common problems of the traditional approach. The key is communication from the market. Material and information is released into the system based on the consumption at the primary control point. Every supplier of raw material as well as every producer along the supply chain is linked to that actual demand. Strategically sized and located buffers of inventory are designed to absorb the unpredictable variability, and sufficient protective capacity is planned to maximize the velocity of the product flow. As a result, the waves of demand are avoided and the productivity of the entire system is made much more predictable. As well, the properly synchronized system is more stable and easier to manage.

12 Partnering with Suppliers

Supplier Relationships

To achieve defect-free manufacturing and fast, dependable customer response, a firm's suppliers must reliably and quickly deliver defect-free materials. To accomplish this, many firms are engaged in supplier development programs. The keys to developing the supplier relationship fully are establishing a relationship based on achieving common goals and using personal relationships to create trust with suppliers. This chapter examines how one electronics company structures its supplier development program to accomplish this.

Supplier Management

Supplier management is a critical factor in achieving quality improvement in many industries. Supplier management or development is the means firms use to develop long-term relationships with their suppliers. Empirical evidence demonstrates that a long-term relationship is critical to mutually improve quality, compress lead time, and reduce costs.

Ten years ago most firms were just beginning continuous quality improvement, and they could make rapid quality improvement by improving their control of their own process without involving suppliers. But, as their internal processes improved, firms began to identify problems caused by defects in their raw material. Often, when firms first realized the costs of poor supplier quality, they reacted by increasing their inspections of incoming material in an effort to improve quality. This immediately improved quality, but it also increased quality appraisal costs. To improve quality and decrease costs simultaneously, companies then initiated supplier development programs.

147

When most supplier development programs began, the initial concern was for a firm to control its suppliers. As stated above, this can be done by requiring inspection of incoming materials. A second step is then to strengthen control by establishing vendor evaluation and certification programs. In these first two stages of supplier development, the firm usually has low trust in its suppliers, so its concentration is on controlling them. In the third stage of supplier development programs, the firm begins to evaluate its vendors and begins to determine which are capable of delivering the required characteristics. During this stage, a firm's trust in its suppliers increases. It is in this stage that firms begin to evaluate suppliers to ensure that the supplier is qualified before doing business. It is here that firms recognize that these evaluations and the selection of the better performing firms will reduce quality costs. These completed evaluations establish the basis for the next stage of supplier development programs, that of vendor certification.

Vendor certification programs allow the purchasing firm to reduce its inspection costs while ensuring that the quality of purchased material remains high. But, to gain the largest improvement in supplier quality, the purchasing firm must completely understand how its purchased materials are used in its own process. This requires the firm not only to understand where the materials are used and what their purpose is, but to understand how much variation there is in the existing product and how this influences or creates variance in the production process. To fully understand this, the firm must first understand how much variation exists in its own processes, so it can separate out the effect of the supplier's variation. It may be that a firm's own process has so much variation that the impact of the supplier's variation is minimal. This stage in the supplier development program often requires that top management in the purchasing firm understand how variance in supplier quality influences quality costs. When all quality costs are tracked and as supplier quality costs became more visible, top managers will emphasize the need for supplier development and begin to request regular reports and presentations to the corporate office about supplier performance.

Different Approaches to Managing the Supplier

Supplier management programs are easy to institute when the purchasing firm represents most of the supplier's business. It is difficult to implement one of these programs if the purchasing firm represents a very small percentage of any supplier's sales, since the firm does not then have the power to dictate improvements to the suppliers. When a purchasing firm has limited power over a supplier, its only option is to select a different approach to managing the supplier. One approach in this situation is to develop personal relationships and use these as leverage to create change at a supplier. A first step in this type of partnering relationships with suppliers is to identify where the supplier and the buyer have shared common improvement goals. One way to do this is to incorporate four components into its supplier development program, which recognize that personal relationships are the key to supplier development.

The first component to build these personal relationships with suppliers is for the firm's staff to understand the supplier's process. To accomplish this, the firm requires a control plan for each material purchased from a supplier. The control plan is a working document that describes the process flow and clearly identifies both for the purchasing firm and for the supplier the critical factors in the supplier's process. This allows the supplier to clearly state what control devices are used and their planned response to out-of-control conditions, and it gives the firm a baseline to measure the supplier's quality improvement.

The control plan requested is often simple (see the example of the form in Figure 12.1). It is important to keep the control plan simple so it that it is a low cost item and so that it is useful as a communication tool. The control plan provides the basis for the purchaser's commodity team to ask questions and identify who are the change makers/process owners in the supplier's organization. These are the individuals who can actually change the process. By understanding who the key players are and which issues are critical to the

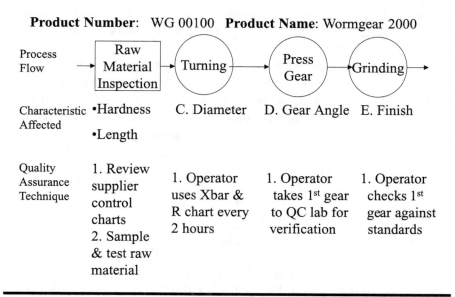

Product Number: WG 00100 **Product Name**: Wormgear 2000

Process Flow	Raw Material Inspection → Turning → Press Gear → Grinding →			
Characteristic Affected	•Hardness •Length	C. Diameter	D. Gear Angle	E. Finish
Quality Assurance Technique	1. Review supplier control charts 2. Sample & test raw material	1. Operator uses Xbar & R chart every 2 hours	1. Operator takes 1st gear to QC lab for verification	1. Operator checks 1st gear against standards

Figure 12.1

supplier's process, the purchaser's commodity team is in position to suggest suitable changes and to support those individuals in the supplier's organization who recognize the need for the improvements that the purchasing firm wants.

A second relationship-building component is personal visits. The purchaser's supplier development team visits at least one supplier a month. They also have supplier representatives come to visit their plant to discuss process improvement corrective actions. This reduces the burden of travel for either group, but allows both groups to remain in close personal contact. This personal contact focuses on the supplier's actual process steps, and it provides the necessary history for open discussions about mutual problems and goals.

Third, to build the trust necessary for the relationship to blossom, the purchasing firm's supplier development team needs to demonstrate constancy of purpose. Suppliers need to perceive that the purchasing firm's actions are always directed toward meeting its own customer's needs even though the techniques used by the purchasing firm may change over time. One method the purchaser uses to do this is to share information with its suppliers about the key characteristics required of its own product for its customers.

Fourth, the purchaser sets the improvement goals jointly with the supplier. By setting the goals jointly, the purchasing firm reduces its control costs, as it does not need to send its own staff to monitor its suppliers. Because it was their improvement goal as well as the purchasing firm's goal, the supplier gathers and reports their own progress. Again, the purchasing firm ensures that all the improvements that it requests of its suppliers are related to the standards desired by their own customers.

These four components are the foundation for a purchasing firm's drive to have long-term partnering relationships with its suppliers. Personal relationships are a key to managing supplier development. They require individuals who can work internally as team members, be able to understand the supplier's process, communicate the supplier's problems back to their own firm, and allow the supplier to own the need to improve. Typical steps in a supplier development program are described below.

Supplier Partnering Process

Usually the supplier development program involves only key suppliers and may actually start with just those suppliers who volunteer. When the supplier starts in the program, it will start to receive regular supplier ratings. If a supplier rating is negative, or if the purchaser rejects material, the supplier is entered into the corrective action stage. This requires the supplier to develop an action plan and certify the improvements. Otherwise, the supplier continuously improves its process.

Supplier Approval Certification

The purchasing firm needs to clearly communicate its expectations to the supplier when it enrolls in its supplier partnering program. This includes clear statements about the evidence that the purchasing firm will use to evaluate the supplier's progress, for example, evidence that a documented quality system is in place is determined by the existence of a quality manual describing the system. A common first-step criteria for the SCIP is that the suppliers have a quality operating system (QOS) in place. A QOS describes the process management system the supplier uses to monitor improvement progress. It contains a central database of key quality indicators and allows the firm to focus on improvement opportunities. A QOS assures the purchaser that the supplier has the basis to achieve quality improvement while decreasing its costs.

Next, each supplier submits the process control plan described earlier to the supplier. The QOS and process control plan allow the purchaser to conduct a quality audit. If there were any unacceptable findings during the audit, the purchasing firm requires the supplier to submit a corrective action plan that identifies: what actions the supplier will take to contain and then eliminate the defect, the coordinator of the plan, and the expected completion date of the correction. The purchasing firm follows up on the supplier's progress through phone contacts with the action plan coordinator and through data submitted by the coordinator.

Supplier Continuous Improvement Program

The purpose of supplier continuous improvement is to establish mutually beneficial, achievable, quality improvement goals. To do this, the supplier must develop the improvement goals. The purchasing firm's role in setting these goals is to provide feedback to the supplier based on its review of the supplier's data, its audit of the supplier, and the key characteristics that its own customers want. Achieving the mutual agreement about the supplier's current status and its goals is crucial to develop trust, decrease control, and move into closer relationships. The purchasing firm's staff members have a major influence on the future relationship at this stage. They need to demonstrate constancy of purpose by using data in their feedback to the supplier and in their analysis. This requires an ability to remain objective and to share information freely.

Basically, the supplier continuous improvement is the Deming cycle (Plan–Do–Check–Act). The supplier sets improvement targets (Plan), takes action to improve the system (Do), collects data and evaluates its progress (Check), and takes action to hold improvement gains (Act) by changing the appropriate procedures. The purchasing firm's role is to support, encourage, and reward those suppliers who actively progress through the cycle. The purchasing firm does this at low cost by tracking and reporting the supplier's progress toward its goals. This progress is tracked primarily using the supplier's own data. The purchasing firm does not need to go back out to the supplier to re-audit to determine its progress. Instead it looks for evidence that the supplier's system is focused on improvement using measurable results. The purchasing firm particularly supports the supplier's efforts to hold its gains by providing recognition of its achievements. The point of supplier continuous improvement is to encourage, monitor, and reward development of high quality materials and services from key suppliers. It is designed so a supplier can achieve realistic, short-term objectives and then repeat the cycle with new objectives.

Supplier Rating

The third subsystem of the supplier partnering process is a supplier rating system. This system gathers data from the supplier's performance in the purchasing firm's own plants. It provides crucial feedback to the supplier. The purchasing firm tracks the quality and delivery performance of the suppliers from period to period. This rating focuses on the quality and delivery performance data. It is a factor in evaluating the supplier status in the continuous improvement cycle. A quarterly report is sent to the supplier, the quality manager, and corporate purchasing. Typically, there are different award levels possible. For example, there might the levels of Quality Member, Quality Partner, and Quality Leader, or A1 Supplier, AAA Supplier, etc. When a supplier achieves a certain level, the real reward is that its products are not inspected and it is given preference in the allocation of long-term business contracts.

Corrective Action

A purchasing firm must submit corrective action reports as required to those suppliers whose quality or delivery performance does not meet the minimum standards of the quality system. This process includes follow-up by the purchasing firm to assure complete resolution of problems. It is important that the purchasing firm encourage suppliers to use a team approach to solving the problem. The use of a team will accelerate the supplier's learning processes. Typically the corrective action report requests a report about the problem-solving process that the supplier is engaged in. An example of a possible corrective action request for problem solving might consists of six basic steps: (1)

identification, (2) request for corrective action, (3) response, (4) follow-up, (5) evaluation, and (6) verification.

The purchasing firm initiates a supplier corrective action request whenever material is rejected during inspection or during the manufacturing process, or if the purchaser discovers a problem during its audit. The supplier submits a corrective action plan describing both containment and improvement actions. The purchasing firm can then use both the supplier's data and its plant's operating data to evaluate the effectiveness of the supplier's improvement. Once the correction is made, the supplier again starts into SCIP. If the supplier does not have a corrective action, it can establish new improvement goals after the supplier rating.

This supplier partnering process is not self-sustaining. It depends on continued input from the purchasing firm's staff members. They need to monitor and measure performance and provide leadership to the suppliers. A central component of each stage in the supplier partnering process is the supplier audit. It is a key factor in the management of the quality system, because it provides data for evaluating and improving system effectiveness. The assumption of an audit is that a quality product can be achieved only through deliberate and conscious acts and/or decisions by the people involved.

Supplier Audit

The audit needs to be a vehicle to develop mutual trust—trust that the purchasing firm has a constancy of purpose, trust that the supplier is capable and is working to improve quality.

The supplier audit is an in-depth examination of the quality system. It examines all elements of the process and the related quality system elements to compare the system with the process specifications. It is a people-oriented activity. It evaluates the acts and/or decisions of individuals regarding a performance standard. It is an important method of providing positive reinforcement to those engaged in the supplier's improvement activities. The supplier audit provides two opportunities: it makes clear the status of the supplier's quality program in respect to the predetermined standards; and it allows the auditor to motivate those responsible to make the necessary changes to correct the shortcomings and to determine the adequacy of the actual changes. Given that the audit's goal is improvement, corrective action requests are not used as punishment, instead they are used to point out areas that are short of the performance standards, to motivate the supplier's improvement.

The quality audit is a communication device. The audit form is given to suppliers in advance. And the form includes descriptions of how each item will be rated. The vital issue is that those items that are audited are important to achieving improved quality.

Many firms model their supplier quality audit on either the ISO 9000 or the Malcolm Baldrige Quality Award (MBQA) criteria. The MBQA has eight categories: leadership, quality planning, information and analysis, raw material quality control, manufacturing process control, human resource use, product quality results, and customer satisfaction.

Costs of Frequent Audits

Supplier audits are a valuable tool in supplier development, but they are also time-consuming and expensive. The purchasing firm's direct costs for its own personnel are high. The direct costs for the supplier are also high as they prepare for and host the audit team. Faced with the quandary that quality audits are an expensive, but very effective tool to improve quality, many firms review their audit process to evaluate how to reduce the costs of their audits while continuing to improve vendor quality. Some firms have found that there is not necessarily any benefit to an annual audit. The first audit gives a picture of the corporate thinking process at the supplier. Given this thinking process, the purchasing firm may identify some suppliers who need an annual audit and others who can be monitored only by the quality of the product they submit.

Some firms allow their mature suppliers to conduct their own self-audit and submit it to them for review. They do reserve the right to inspect and verify the accuracy of the audit later, but usually they do not need to do so. These firms may also ask the supplier to conduct product audits during the year and compare the actual process to their documented procedures and the results to their targets. The key to this self-audit is the purchasing firm's personal relationships with the supplier and its ability to track the supplier data. The purchasing firm can interpret the data because the commodity team previously analyzed the supplier's process control plan to understand the importance of the data from the supplier.

Supplier Involvement

As part of supply chain management, many firms are transferring some responsibility for innovation to their suppliers. These firms expect the suppliers to have special expertise in their given product areas.

Early supplier involvement (ESI) is a technique to transfer responsibility for product innovation to the supplier. This transfer requires a large amount of time on the part of the purchasing firm. This means that the firm must focus on transferring responsibility to those firms that are the most likely to produce a good payoff for the effort. One technique to do this is to segment all parts that are purchased into three categories.

- Standard parts that require no customization (e.g., nuts and bolts)
- Black-Box parts where development has always been done by the supplier
- Custom-made parts which are currently designed by the purchasing firm.

It is in the custom-made parts area that there is the most potential of supplier innovation. It is here that the purchasing firm wants to move from the status of supplying the design. A first step is for the purchasing firm to share the design and ask for suggestions, but still retain full responsibility for the design. The next step may be to source the design, where the supplier takes full responsibility for a system, incorporating parts also designed by the supplier.

Successful implementation of ESI requires that the firm carefully prepare to align its strategy, human resource practices, and operating policies. The second phase of ESI is to select a partner and negotiate a new kind of supply arrangement. The arrangement must address the need for confidentiality, pricing, communication, and mutual recognition. The third phase of the process is the operations phase. Here there must be a growth in trust between the partners. This is accomplished by understanding each other's processes and acknowledging altered time requirements.

13 Quality Management

The basic tools for quality management are discussed in Chapter 5. Some additional tools such as Quality Function Deployment (QFD) and Failure Mode Effect Analysis (FMEA) are discussed in this chapter. In addition, quality management systems such as ISO 9000, QS 9000, and systems based on the Malcolm Baldrige National Quality Award (MBNQA) are explained in the first part of the chapter. In the second part of the chapter, techniques to improve quality throughout the entire supply chain are discussed.

Quality—Conformance to requirements or fitness for use. Quality can be defined through five principal approaches: (1) Transcendent quality is an ideal, a condition of excellence. (2) Product-based quality is based on product attribute. (3) User-based quality is fitness for use. (4) Manufacturing-based quality is conformance to requirements. (5) Value-based quality is the degree of excellence at an acceptable price. Also, quality has two major components: (1) quality of conformance—quality is defined by the absence of defects, and (2) quality of design—quality is measured by the degree of customer satisfaction with a product's characteristics and features.

APICS Dictionary, 9th edition, 1998

To be effective, a quality management system must be based on a thorough understanding of customer needs, what is required to meet those customer needs, and a control system to ensure that the firm is actually performing as required.

Every business is trying to satisfy a need. Managers must know what that need is and share this with the employees. The customers of the business are those people or firms whose needs are being satisfied. For example, the customers of a small retail paint store could have needs not only for the physical products (i.e., paint, paint brushes, tape, etc.) but also for information about products and techniques useful in solving tough painting problems.

To satisfy its customer's needs firms must fulfill specific requirements. These requirements might include technology, technical and process knowledge, support equipment and services, and customer training. Customers of a small paint store who want knowledgeable advice as part of the product/service bundle that they purchase from the store create a requirement that the store staff have a high level of expertise in the use and application of different paints in different types of situations.

Finally, a firm must have a control system in place to ensure that it does what it said it would do. For example, if the customer requested 100 units of a certain product on Friday, and the firm does not deliver until Saturday then the transaction was not fulfilled. Or, if a paint store advertises that it can mix its paint to match any paint chip brought into it, it needs to have a system in place that allows it to track how closely it can match paint colors and how often it fails to match paint colors.

The creation of a control system requires that the firm create or establish quality specifications. The lack of a clear quality specification means that the workers do not precisely know what a quality job looks like. Once the specifications exist, it is important that they be shared with all employees. For example, if the paint store employees do not understand that they are expected to "play" with the tints until they match a customer's paint chip exactly, they might give the customer paint which is mixed so that it is only close to the exact color the customer wanted. The quality specification is necessary for the employees to understand when they have done the job correctly.

Specification—A clear, complete, and accurate statement of the technical requirements of a material, an item, or a service, and of the procedure to determine if the requirements are met.

APICS Dictionary, 9th edition, 1998

Quality management requires the same basic management functions as required for any other aspect of the business. It requires managers to plan, organize, lead, and control. A portion of any business plan must include how to create a culture that values quality and management systems that ensure quality. Given a quality management plan, a quality assurance system needs to be put into place to provide the control necessary to ensure that the quality system requirements are fulfilled. The quality assurance system consists of the quality policies, rules, and system in place. The quality assurance system specifies which quality control techniques are to be used. These are the techniques used on the shop floor to ensure that the requirements of the quality system are being met. This includes the use of statistical process control.

Quality assurance/control—Two terms that have many interpretations because of the multiple definitions for the words "assurance" and "control." For example, "assurance" can mean the act of giving confidence, the state of being certain, or the act of making certain; "control" can mean an evaluation to indicate needed corrective responses, the act of guiding, or the state of a process in which the variability is attributable to a constant system of chance causes. One definition of quality assurance is all the planned and systematic activities implemented within the quality system that can be demonstrated to provide confidence that a good or service will fulfill requirements for quality. One definition for quality control is the operational techniques and activities used to fulfill requirements for quality. Often, however, quality assurance and quality control are used interchangeably, referring to the actions performed to ensure the quality of a good, service, or process.

APICS Dictionary, 9th edition, 1998

This chapter is primarily concerned with advanced quality planning and control tools used in the quality management systems. The basic tools were discussed in earlier chapters.

Quality Function Deployment (QFD)

Quality function deployment (QFD) is a tool that is used to specify all major requirements of customers and then evaluate how well the designs meet or exceed those requirements. It is used in developing new products, improving existing products, and developing processes to manufacture the products. QFD is a set of methods to take all of the information gathered from a firm's customers and potential customers and organize it to facilitate the product development process. QFD is a communication and translation tool. It allows a cross-functional team to share information effectively.

Quality function deployment (QFD)—A methodology designed to ensure that all the major requirements of the customer are identified and subsequently met or exceeded through the resulting product design process and the design and operation of the supporting production management system. QFD can be viewed as a set of communication and translation tools. QFD tries to eliminate the gap between what the customer wants in a new product and what the product is capable of delivering. QFD often leads to a clear identification of the major requirements of the customers. These expectations are referred to as the voice of the customer (VOC).

APICS Dictionary, 9th edition, 1998

Many people report that the house of quality (i.e., QFD) appears intimidating when they first see it. And, at first glance it does appear hopelessly crowded. It is actually easy to understand if you work through each of the building blocks individually. The house of quality (notice the roof in Figure 13.1) is built section by section. It is not possible to build all the sections at once. In fact, it is probably impossible to build sections simultaneously.

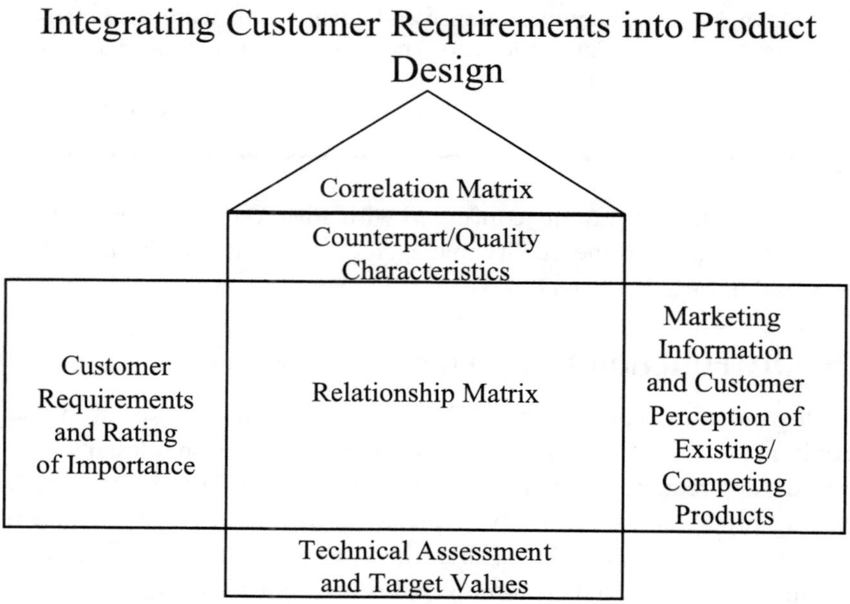

Integrating Customer Requirements into Product Design

Figure 13.1 House of Quality—Integrating Customer Requirements into Product Design

The purpose of a QFD is to facilitate cross-functional communication. So, a QFD is built using a team. A QFD is a planning tool and even the simple products require a cross-functional team to plan them correctly. All relevant areas of the firm involved in producing the product need to be represented. Often this includes appropriate personnel from purchasing, marketing, production operations, design, and process engineering.

The first step is to identify the customer requirements. At this step it is usually marketing's job to represent the voice of the customer. Marketing must determine what attributes the customer really wants from the product or service. Most customers will give only general, vague, difficult to implement statements about the products or services they may want to purchase. For example, in response to a question about what a patron expects from a hair salon, the patron might respond that he/she wants a "good haircut, at a fair price, delivered in comfortable surroundings." The challenge is to get more information from the customer without distorting it by putting words into his/her mouth. This requires those interacting with the customer to ask follow-up questions. For example, "How do you know when a haircut is good?" "What is a fair price?" "What makes the surroundings comfortable?" This can sometimes be facilitated by showing the customer images or by having a prototype product for him/her to examine. Once this information is obtained, it is listed as a set of customer requirements in a simple single column table as shown in Figure 13.2. This column just lists *what* the customer wants in the product or service.

- **Answer the question of what attributes the customer desires from the product or service**
  - general, vague, difficult to implement
  - example *friendly staff at bank*

WHAT does the customer want

Want 1
Want 2
Want 3

Figure 13.2   Identifying the Customer Requirements

● List one or more 'HOWS' for each 'what"

● Each 'How' translates a 'what' into reality

    – example. *We create the perception of friendly staff by having our staff memorize and use the customer's name, etc.*

WHAT does the customer want?   HOW do we provide each what?

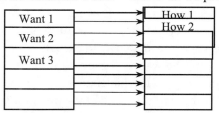

Figure 13.3   Identifying Product Characteristics

The next step is to identify *how* to provide what the customer wants. This is a technical characteristic. For example, if a customer wants a cup of coffee served scalding hot, then how the firm does this is by serving coffee at 200°F. To be able to serve it scalding hot the firm might also have a second *how* that states that coffee will be made using water at a minimum of 211°F. Identifying and listing these *hows* is also a group process. However, this portion of development may be dominated by the operations and engineering representatives, because it requires some technical understanding of how the product is manufactured and/or the service is delivered. As a rule of thumb, the team developing the QFD attempts to identify at least one or two *hows* for each *what* as shown in Figure 13.3.

For a complex product, the lists of *whats* and *hows* produced by an active QFD group could now be very tangled. That is, 2 or 3 *whats* are being satisfied by the same *how*. It is also highly likely that group members will notice that some of the *hows* contradict each other. For example, in a bank, it may be that one *how* identified is to have each teller memorize the names of all the regular customers and to then use the name when speaking to the customer. It is possible that a second *how* is to limit the interaction with the customer to less than a minute total. Some group members might object that these contradict each other, given the desire of many customers to respond to friendliness by asking questions or chatting. These possible contradictions are noted, but no other action is taken at this stage.

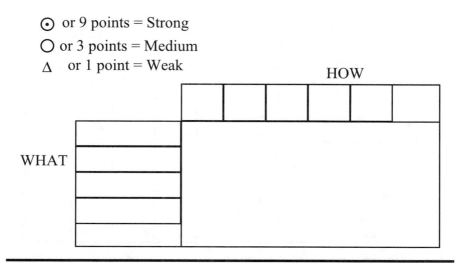

Figure 13.4   Define Relationships between What and How

Once all of the *hows* for each of the *whats* have been listed, the relationships between the *hows* and the *whats* are then identified. This is done by taking the 2 columns and rotating the *how* column 90° up from the *whats* as shown in Figure 13.4. Once the *hows* are on top and the *whats* are on the side, the relationship between the *whats* and the *hows* is identified as being either a strong relationship (9 points), a medium relationship (3 points), or a weak relationship (1 point). Some people use symbols instead of assigning points. This alternative is also shown in Figure 13.4. If there is no relationship between a *what* and a *how* the intersection is left blank.

At this stage it is useful to check for blank rows or columns. A blank row means that no *hows* have been identified for a particular *what.* The team needs to revisit the *what* that has been neglected and determine if it is valuable enough to identify some *hows* for it. If it is not important enough to go back and create some *hows* then it can be erased. If there are some blank columns, then it means that some *hows* have been created that do not help to create any of the *whats* that the customer wants. These blank columns are erased. This is illustrated in Figure 13.5. Notice that the second column from the right side has no relationship to any of the *whats* in the left column. So, this *how* will be erased. It is important to edit the QFD as it is being created, to ensure that the complexity does not overwhelm the team developing the QFD.

Be sure to note in Figure 13.5 that many of the *whats* are related to more than one of the *hows*. This is also true for the many of the *hows*. The next step is to determine *how much*. The *hows* state the technique or material that will be added to a product to produce a desired *what,* and the *how much* portion

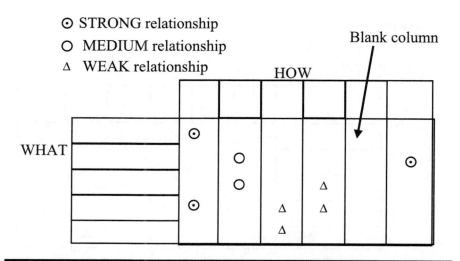

Figure 13.5   Editing the QFD

of the QFD states the specific amount that is needed to provide the *whats* the customer wants. For example, if the manufacturer of refrigerators has found that customers like the door of the refrigerator to feel strong, the manufacturer may have decided to produce the refrigerator door out of rolled steel. The *how much* block for the rolled steel would specify the type of steel and the thickness of the steel to be used. As shown in Figure 13.6, the *how much* blocks are put at the bottom of the relationship matrix. The *how much* rows provide an objec-

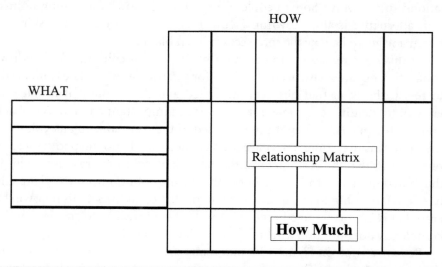

Figure 13.6   How Much

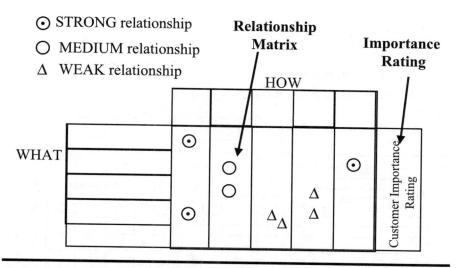

Figure 13.7 Customer Importance Rating

tive measurement of whether the customers' *whats* will be satisfied by a given *how*. The *how much* row also identifies targets for further development.

At this point the basics of the QFD have been created. The basic QFD shows what has to be done, how it has to be done, how much is to be done, and the relationships of what is to be done to how it is to be done. These elements form the essential QFD. There are many extensions to the QFD, but only the most common of these extensions will be examined here.

The first extension to the QFD tries to quantify the importance of each of the *whats* identified by the customer. This data can be based on subjective feelings, or it can be based on more subjective impressions by marketing as described in Figure 13.7. It is of course much better to have quantitative data about how customers evaluate the importance of a particular product or service attribute, but it is possible to proceed without this information.

This importance rating is placed on either the left or right side of the relationship matrix. In the example shown in Figure 13.8, it is shown in the column immediately to the right of the customer requirements (i.e., *what*). This importance rating can be either a subjective estimate of how important a customer considers a particular *what*, or it can be a quantitative estimate. If it is not a quantitative estimate, then it should be made by someone who has an intimate understanding of the customer. This will be used to make crucial decisions in the product and process design, so it is important that this information be correct. A rating of 1 means that the *what* in question is unimportant and a rating of 5 means that it is extremely important.

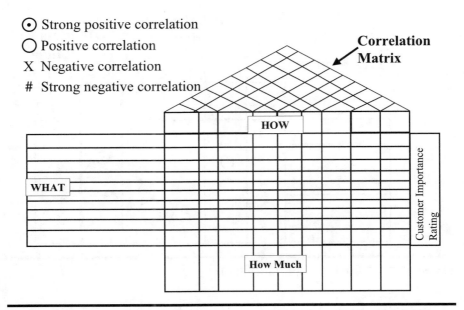

Figure 13.8   Correlation Matrix

Another extension to the QFD is to examine the relationship of the *hows* to each other. We are interested in both negative and positive relationships between the *hows*. This is shown in Figure 13.8 as a correlation matrix. The correlation matrix shows the degree to which each *how* is correlated to every other *how*. The correlation matrix is the triangle on top of the relationship matrix. It forms the roof of the QFD.

Strong positive correlations (shown by a circle with a solid circle inside of it) show that one *how* strongly supports another *how*. A positive correlation is shown by an empty circle. Positive correlations indicate where it is possible to achieve resource efficiencies. Negative correlations also exist. A negative correlation shows the adverse effects of one *how* on another *how*. Negative correlations are shown in the QFD with an *X*, while strong negative correlations are shown with a pound sign (#).

For our purposes this is enough development of the QFD. The development of the QFD to this point leaves us with a QFD that shows what the customer wants and the specific characteristics of the service or product that we will include in the design to provide what the customer wants. In addition, it tells how much of the service or product will be provided to satisfy what the customer wants. The relationships between what the customer wants and how we will provide it is shown and the relationships between the different methods of how we will meet the customer's wants is also provided. By incorporating the customer importance ratings into the QFD we have information that allows us to make decisions about trade-offs. The QFD can be used to share information with our suppliers about the importance of certain supply characteristics to the final customer. It helps the supplier understand why the products they are supplying to us must meet certain criteria. It is also a vehicle to involve the supplier in product development. The supplier could help develop the portion of the QFD that is related to the product they supply. The supplier could also use the QFD to examine their own processes and the products that they deliver.

ISO 9000 Quality Management System

In 1987 the International Organization for Standardization created a set of standards for Quality Management Systems (QMS). This group of standards was given the numbers 9000, 9001, 9002, 9003, and 9004. The standards were revised in 1994, and another revision will be published in 2000. The final standards were not published at the time of this book's publication, but there are several key differences between ISO 9000 (1994) and ISO 9000 (2000). These differences will be addressed as the standards are discussed.

When trying to understand the standards and their revision process, it helps to understand the organization that creates these standards. The International Organization for Standardization is based in Geneva, Switzerland. Its purpose is to promote international trade. To facilitate international trade it creates standards for a wide variety of products and processes. For example, there is an ISO standard for film speed (e.g., see the ISO number on a packet of film). ISO is not a government organization, but every country can send a representative. The American National Standards Institute (ANSI) is the official U.S. representative to ANSI. All the ISO standards are voluntary. There is no legal requirement for a firm to become registered to these standards. Most companies report that they seek registration because their customers require it.

All the ISO standards are developed using the same basic process. A technical committee is formed to consider a standard. Every country can send a representative to this technical committee. In the United States, the representatives are often from those firms which will be impacted by the proposed new standard. The representatives on this technical subcommittee subdivide the task of developing a standard into working groups. Over a period of years the working groups report back with drafts of the portions of the standard they were responsible for developing. These drafts are then considered by the entire technical committee and, once consensus is reached, there is a vote.

Remember, that the ISO standards are voluntary in the sense that they are not required by government regulation. However, many firms are forced to implement ISO 9000 because their customers require them to do it. When

ISO 9000 was issued in 1987, few U.S. firms elected to become registered to the standard. But, as international firms recognized the advantage of adhering to the standard they began to require their suppliers to be registered to the standard. These suppliers in turn required it of their suppliers. As of the end of 1998, there were over 200,000 firms in the world who had been registered to the ISO 9000 standards.

In the United States the administration of ISO 9000 is conducted jointly by ANSI and the American Society for Quality (ASQ). They certify registrars who send auditors to companies that want to become registered or certified to the standards.

Although the new ISO 9000:2000 standards have been published, there is a transition period during which firms can retain their certification to the ISO 9000:1994 standards, so it is important to understand some basics about the 1994 standards. Therefore, both ISO 9000:1994 and ISO 9000:2000 will be discussed here.

In the 1994 standards, only ISO 9001, 9002, and 9003 are actually the quality management standards to which a firm can be certified as adhering. ISO 9001 gives guidelines for firms to help them determine which standard to adopt, and it contains definitions of the terms used in the standards. ISO 9004 is not a standard; rather, it is a set of guidelines to help firms implement the standards. In the 2000 standards there are only 3 standards in the ISO 9000 family. These will be:

- ISO 9000:2000 (Fundamentals and vocabulary)
- ISO 9001:2000 (Requirements)
- ISO 9004:2000 (Guidance for performance improvement)

The ISO 9001:2000 has only the one quality management system standard, replacing the current set of 9001, 9002, and 9003. The ISO 9001:2000 and ISO 9004:2000 standards will have consistent numbering and will be compatible with the environmental management standards (ISO 14001).

ISO 9000 series standards—A set of five individual but related international standards on quality management and quality assurance developed to help companies effectively document the quality system elements to be implemented to maintain an efficient quality system. The standards, initially published in 1987, are not specific to any particular industry, goods or service. The standards were developed by. . . a specialized international agency for standardization composed of the national standards bodies of 91 countries.

APICS Dictionary, 8th edition, 1998

It is important to recognize that the ISO 9000 standards are not product standards. They are standards about the documentation of the quality system. This means that they are generic standards which can be applied to any company in any industry. In the ISO 9000:1994 standards, the difference between ISO 9001, ISO 9002, and ISO 9003 is the amount of work that a company does internally on a product. A company that designs, manufactures, and tests its product would be certified to ISO 9001. A company that only manufactures products to designs given to it by someone else is certified to ISO 9002. A company that only tests and distributes product would be certified to ISO 9003. As an example, a firm might have a design center at one plant. This design center might produce designs for all of the firm's plants, but only the one site with the design center would need to be certified to ISO 9001. The other plants in the firm would be certified to ISO 9002, because they do not manufacture product. This distinction will not be made with the ISO 9000:2000 standards.

Given what we have established earlier about quality management, it is clear that ISO 9000 will help firms achieve a quality product. For example, earlier we stated that meeting customer requirements is a basic need to provide quality. Item 3 of ISO 9001 is Sales Order Review (see Table 13.1).

Table 13.1   Elements of ISO 9001:1994

Element
1. Management Responsibility
2. Quality System
3. Sales Order Review
4. Document Control
5. Design Control
6. Purchasing
7. Customer Supplied Product
8. Product Identification/Traceability
9. Process Control
10. Inspection and Testing
11. Inspection, Measurement, and Test Equipment
12. Inspection and Test Status
13. Control of Non-conforming Product
14. Corrective action
15. Handling, Packaging, Storage, and Delivery
16. Quality Records
17. Internal Quality Audits
18. Training
19. Service
20. Statistical Techniques

When it is effective in improving quality in a manufacturing system, it is because the ISO 9000:1994 documents accurately capture the procedures of the process. And further, it is because the employees actually use the procedures specified in the ISO 9000 documents. The purpose of the revisions in the ISO 9000:2000 is to allow organizations to focus on their major processes and emphasize the customer satisfaction. In the ISO 9000: 2000 revision, these 20 elements are contained in 4 clauses as shown in Table 13.2.

One major change introduced by the ISO 9000:2000 standard is the emphasis on the customer. It was always possible for companies to emphasize customer satisfaction under the old standards, but it was not required. Another significant change in ISO 9000:2000 is that the standards require that managing change is included in the planning for quality.

Quality managers of companies that have installed ISO 9000 frequently report that they should have put a system like this in place earlier. It is often the

Table 13.2   ISO 9001:2000 Clauses

5. Management Responsibility
   5.1 Management Commitment
   5.2 Customer Focus
   5.3 Quality Policy
   5.4 Planning
   5.5 Administration
6. Resource Management
   6.1 Provision of Resources
   6.2 Human Resources
   6.3 Facilities
   6.4 Work Environment
7. Product Realization
   7.1 Planning of Realization Processes
   7.2 Customer Related Process
   7.3 Design and/or Development
   7.4 Purchasing
   7.5 Production and Service Operation
   7.6 Control of Measurement Devices
8. Measurement, Analysis, and Improvement
   8.1 Planning
   8.2 Measurement and Monitoring of Product
   8.3 Control of Nonconformity
   8.4 Analysis of Data
   8.5 Improvement

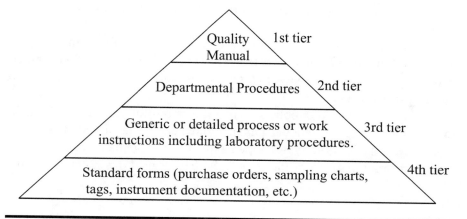

Figure 13.9 The Pyramid of Quality (Reprinted from Lamprecht, p. 32 by courtesy of Marcel Dekker, Inc.)

first time that they have organized into one place all of the procedures and systems that were being used to ensure quality. When these are gathered together, problems are often revealed which can then be addressed.

There is a quasi-standard approach to installing an ISO 9000 system in a firm. It is not required by the standards, but is used by a large number of firms. This approach is to organize the quality system in terms of the pyramid illustrated in Figure 13.9. This visual model also illustrates the amount of detail, which is to be found at each level of the ISO 9000 system. At the top level, the Quality Manual, no detail is provided; rather, the purpose of the top level is to state the firm's policy in regards to each section of the ISO 9000 standards. This is often kept very simple and short. The second level of the system requires documenting the departmental procedures for each relevant section of the quality standards. This can be done simply with a flow chart, or with text. The third level of the pyramid can become very detailed, as it includes the work instructions. The final level in the pyramid is the documentation that the higher 3 levels have been done. This might consist of the SPC charts that the manual states that employees will develop for certain processes. These can be stored as paper in files, or they can be stored electronically.

QS 9000

After ISO 9000 was established, the U.S. auto companies decided that it did not go far enough. They wanted their suppliers to have a quality management system in place that not only described what the company did, but would ensure that the supplier was performing certain practices in a specified manner.

Remember from above, that ISO 9000 is voluntary. A firm can have poor practices, but if they are documented, the firm can become ISO 9000 certified. However, they would have to disclose in their documentation that they have poor quality practices.

To ensure that their suppliers all met a certain minimum level of quality management practices, the Big Three automakers in the United States (GM, Ford, and Chrysler) jointly developed a quality management system that exceeded the requirements of ISO 9000:1994. They called this system QS 9000.

QS 9000—Quality management system requirements cooperatively developed and adopted by the "Big Three" automobile manufacturers, Chrysler, Ford, and General Motors, along with certain truck manufacturers. QS-9000 incorporates all of the main elements of the ISO 9001 standard and describes the minimum quality system requirements to emphasize continuous improvements, defect prevention, consistency, and elimination of waste.

APICS Dictionary, 9th edition, 1998

One important difference between ISO 9000 and QS 9000 is that QS 9000 is not a voluntary standard. The auto companies require that all of their Tier 1 suppliers be certified to QS 9000. QS 9000 is more extensive than ISO 9000. It requires additional quality system elements and it includes advanced quality planning. In addition, it requires that control and quality plans be developed. For example, QS 9000 requires the development of Failure Mode Effect Analysis (FMEA). The similarities and differences between ISO-9000 and QS 9000 are shown below in Table 13.3.

The QS 9000 system has 20 sections which are fully compliant with ISO 9000. The differences in these 20 sections is that QS 9000 QS requires more than ISO does. In these sections, QS will stipulate how something is to be done, not only that it is to be done as ISO 9000 does. In addition, QS has two additional sections with requirements that are not in ISO 9000. These are shown in Table 13.3 as Section 2 and Customer-Specific Requirements.

Most firms have the overall design of their QS system organized in the same manner that they had their ISO 9000 quality management system organized. This allows them to obtain both the ISO and QS certification at the same time. This is shown in part in Table 13.3.

Table 13.3 Comparison of QS 9000 to ISO 9000:1994 Elements

QS-9000 Title		ISO Element	ISO Compliance
4.1	Management Responsibility	4.1	Full
4.2	Quality System	4.2	Full
4.3	Contract Review	4.3	Full
4.4	Design Control	4.4	Full
4.5	Document and Data Control	4.5	Full
4.6	Purchasing	4.6	Full
4.7	Control of Customer Supplied Product	4.7	Full
4.8	Product Identification and Traceability	4.8	Full
4.9	Process Control	4.9	Full
4.10	Inspection and Testing	4.10	Full
4.11	Control of Inspection, Measurement, and Test Equipment	4.11	Full
4.12	Inspection and Test Status	4.12	Full
4.13	Control of Non-Conforming Product	4.13	Full
4.14	Corrective and Preventive Action	4.14	Full
4.15	Handling, Storage, Packaging, and Delivery	4.15	Full
4.16	Control of Quality Records	4.16	Full
4.17	Internal Quality Audits	4.17	Full
4.18	Training	4.18	Full
4.19	Servicing	4.19	Full
4.20	Statistical Techniques	4.20	Full
Section 2			
Production Part Approval Process		—	—
Continuous Improvement		—	—
Manufacturing Capabilities		—	—
Customer-Specific Requirements		—	—
Chrysler-Specific Requirements		—	—
Ford-Specific Requirements		—	—
General Motors-Specific Requirements		—	—
Truck Manufacturers—Specific Requirements		—	—

Malcolm Baldrige National Quality Award

The Malcolm Baldrige National Quality Award (MBNQA) was created to promote quality and to recognize those companies that excelled in quality. The award criteria and performance measurements were developed through extensive consultation with practitioners and consultants. These criteria are periodically updated as more information about how to achieve high quality levels is developed.

Table 13.4   Malcolm Baldrige Criteria for Performance Excellence

1. Leadership
2. Strategic Planning
3. Customer and Market Focus
4. Information and Analysis
5. Human Resource Focus
6. Process Management
7. Business Results

Currently the award is given annually to 3 companies. One award goes to a small company, another to a large company, and a third to a service company.

The criteria for the award are shown in Table 13.4. The MBNQA is not examined in depth here, because most firms use quality management systems that are based on ISO 9000.

Whether the firm uses a quality management system that is based on ISO 9000 or one based on the MBNQA, firms which are practicing total quality management attempt to involve their employees in the effort for improvement. To do this, they develop and use standard problem-solving methods. One standard problem-solving method is discussed below.

Standard Problem-Solving Methods

Many firms have adopted a standard problem-solving approach which managers and employees alike are expected to use. This helps create the quality culture that most quality experts believe is necessary. A standard problem-solving method also encourages communication across all of the functions and accelerates learning throughout the organization. At the same time, using a standard method that is based on the scientific method leads to improved problem solving, because problem solving requires communication and the use of the scientific method.

The use of standard problem-solving methods is encouraged by ISO 9000: 1994, which states in element 4.14 that "The supplier shall establish and maintain documented procedures for implementing corrective and preventive action." In that same section, QS 9000 also includes the requirement that the "supplier shall use disciplined problem solving methods when an internal or external nonconformance to specification or requirement occurs."

Standard problem-solving techniques are used not only to correct defects but also to improve existing processing. In today's environment problem solving cannot be left to only the company genius. By creating a standardized

Table 13.5 Standard Problem-Solving Example

Step	Description
1	Select Theme
2	Collect and Analyze Data
3	Analyze Causes
4	Plan and Implement Solution
5	Evaluate Effects
6	Standardize Solution
7	Reflect on Process (and next problem)

Adapted from Shiba, Graham, and Walden, 1993.

problem-solving method it is possible to involve everyone in solving problems. The exact problem-solving steps vary from firm to firm, but a common example is referred to as the 7 Step Method. The 7 steps are listed in Table 13.5.

Step 1 is to select the theme. This is important because a problem which is too difficult will discourage further efforts and a problem that is trivial will ruin the team's morale. The problem needs to be one that repeats and is related to either the internal or external customer.

Step 2 is to collect and analyze data. Here, it is important that real facts be gathered and that these be analyzed using graphs and the other tools discussed above. The important point here is that the team logically explore the problem with data.

Step 3, analyze causes, is an examination of the causes of the problem identified in Step 2. This usually uses the cause-and-effect diagram. It is important that the CED be developed in depth. It must be clear how the team reached their conclusions about the causes of the problem.

Step 4, plan and implement solution, is a statement of how the team intends to solve the problem. It is important that the team understand the limits of its authority, so that it will develop local solutions that are very focused on the root cause(s) identified in Step 3. They are not authorized to change the entire system. The team needs to analyze the problem and identify different alternatives to solving the problem. It is valuable for the team to assign the tasks of implementing the solution using the 4 Ws and 1 H: who, when, where, what, and how.

Step 5, evaluate effects, is the step at which the team confirms whether the improvement that was sought was actually achieved. This can be illustrated by gathering data and analyzing it as was done in Step 2.

Step 6, standardize solution, identifies the methods to ensure that the improvement is permanent. When the people leave, will the improvement stay?

Step 7, reflect on the process (and next problem), is a self-evaluation by the team. How could they have performed the 7 steps better? What did they learn by doing this?

Of course this method will only help to create a quality culture if everyone uses it and demonstrates that he/she is using it to solve problems. In many companies this is encouraged by having the teams that use the method share with other teams and managers what they did at each step and the results of their work.

14 | Work Teams

eams are the method more and more firms are using to manage their production process. A team is typically thought of as a group of individuals who are working together for a particular goal. There are multiple variations on the types of teams in firms. There are problem-solving/decision-making teams, production teams, self-directed work teams, semi-autonomous work teams, engineering or project teams, and management teams. This chapter examines two concepts about teams. First, it examines two primary types of teams—cross-functional and self-directed work teams. Second, this chapter examines the concept of empowerment.

Cross-Functional Work Teams

A cross-functional team is one whose members are drawn from different departments to share their expertise. For example, purchasing may set up a cross-functional team to decide where to purchase some new piece of equipment. The team members in this situation would include a representative from maintenance, finance or accounting, production planning, process engineering, and purchasing.

Cross-functional work teams are formed when there is a specific task to be completed, which requires expertise that is contained within many functions of the organization. For example, a quality improvement team might be formed to reduce defects in a particular process. The members of this team might include a production worker, a product designer, a process engineer, a sales representative, and someone from purchasing. If the team is not given a clearly defined task initially, its first order of business is to define the task and assign responsibilities. Typically these employees will continue to be responsible for their regular work as well as participating on the team.

Cross-functional team—A set of individuals from various departments assigned a specific task such as implementing new computer software.

APICS Dictionary, 9th edition, 1998

In a cross-functional team, one person assumes the role of the team leader, who is responsible for ensuring that the work is accomplished, scheduling the meetings, setting the agenda, and conducting follow-up with all members to ensure that the team has met its goals. The role of the leader is crucial in a cross-functional team.

The purpose of the meetings is to communicate among the team members, to discuss findings, and then to decide on the next step. For example, if a cross-functional team is assigned to shorten the lead time of a particular product from 6 months to 3 months, the team would meet to outline the tasks and assign responsibilities. The first step might be the creation of a flow chart that includes the time spent at each stage of the process. Each member of the team would help gather the data. When this information is compiled, the team might decide to investigate one particular set of steps in the process because they consume the most time. The team would then assign responsibilities to individual team members to obtain data about those stages. Once that information was obtained the team would meet again and discuss the data. The team would then decide when to meet again. This cycle of meeting and working outside of the team meetings would continue until the problem was solved.

Self-Directed Work Team

In this book, the terms *work teams, semi-autonomous teams, autonomous teams, self-directing teams,* and *self-managing teams* are all considered to be examples of self-directed work teams. A self-directed work team is a group that routinely performs the same set of tasks together. The group members are often a production team. For example, the team may be responsible for the production of molded rubber V belts from raw rubber stock. Generally, the autonomous team is given a wide range of tasks, including planning tasks, and is held responsible for the performance of these tasks.

Self-directed work team—Generally, a small, independent, self-organized, and self-controlling group in which members flexibly plan, organize, determine, and manage their duties and actions, as well as perform many other supportive functions. It may work without immediate supervision and can often have authority to select, hire, promote, or discharge its members.

APICS Dictionary, 9th edition, 1998

Self-directed work teams are given authority to solve problems related to their jobs. The amount of authority that is given to a team depends on the company. For example, some companies allow the teams to interview job candidates for a position on the team and either to make the final decision or to have input into the final decision about hiring.

Many companies already organize their shop floors with teams. The teams are usually trained in total quality tools and techniques. For a self-directed work team to function well, it has to have a clear structure to provide a framework for decision making. A common structure for this type of team is referred to as either the spoke or star structure. This means that the team's responsibilities are classified into distinct areas. Typically, one team member is appointed to fulfill the role of leader for each responsibility area. Common star leadership positions include scheduling, human resources, quality, and safety.

A common illustration of the star concept, which is used to train team members, is shown in Figure 14.1. For example, an individual may be the scheduling leader of Team ABC during 1998 and then become the quality leader of Team ABC during 1999, while someone else will become ABC's scheduling leader in 1999. This offers several advantages to a firm. First, it reduces costs as supervisors are not needed. Second, it creates a clear structure for the team and provides a clear communication channel between management and the team regarding each issue the team is responsible for. A third advantage of the structure is that the team members clearly understand what their responsibilities and authority are in each star or spoke area. A fourth advantage is that this team structure allows interteam communication about shared problems in the process. Development of a structure by which teams can assume responsibility for waste minimization is a crucial step. Research suggests that the design of a team is the most critical element in its success. Specifically, it is possible that the design of the team is more important than the quality and level of coaching or facilitation that the team receives from management (Wageman, 1997).

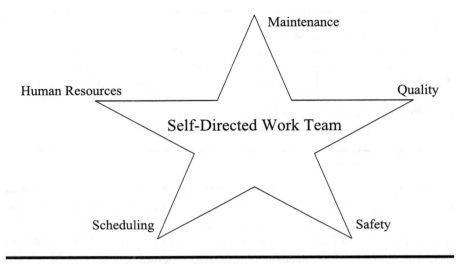

Figure 14.1 Example of "Star Point" Team System

Empowerment

Teams are often used to empower employees. Empowerment is a means of involving every employee in the improvement process. The purpose of this is to make improvement everyone's responsibility. A company which empowers its workers recognizes that they can make improvements faster if everyone who works for the firm is contributing to the improvement efforts.

> **Empowerment**—A condition whereby employees have the authority to make decisions and take action in their work areas without prior approval. For example, an operator can stop a production process if a problem is detected, or a customer service representative can send out a replacement product if a customer calls with a problem.
> *APICS Dictionary,* 9th edition, 1998

Typically, empowerment consists of giving a line employee the authority to make decisions about his or her work environment. This may include determining the work schedule, assigning tasks among team members, or conducting the mundane administrative tasks of operating a plant, such as completing leave forms. This usually occurs within the team, which provides a structure for decision making and goal setting.

Employee empowerment—The practice of giving nonmanagerial employees the responsibility and the power to make decisions regarding their jobs or tasks. It is associated with the practice of transfer of managerial responsibility to the employee. Empowerment allows the employee to take on responsibility for tasks normally associated with staff specialists. Examples include allowing the employee to make scheduling, quality, process design, or purchasing decisions.

APICS Dictionary, 9th edition, 1998

Empowering employees requires more than just telling them to go and make their own decisions. It requires that managers create an infrastructure of organizational policies and practices which supports employee decision making or empowerment. It is through the creation of this infrastructure that management creates an empowered work environment. A key component of this empowered work environment is that employees understand what decisions they can and cannot make and further that they understand what information they must have to make good decisions.

Empowerment is often viewed as one step beyond employee involvement. Employee involvement is usually perceived as participation of the employees in some or all of the decision-making process, but without having the authority to make the decision themselves. The difference between employee involvement and employee empowerment is that the empowered employee has more authority or power to make decisions.

Employee involvement (EI)—The concept of using the experience, creative energy, and intelligence of all employees by treating them with respect, keeping them informed, and including them and their ideas in decision-making processes appropriate to their areas of expertise. Employee involvement focuses on quality and productivity improvement.

APICS Dictionary, 9th edition, 1998

Benefits of Teams

One major benefit of teams is that they reduce costs by eliminating some of the need for middle managers. But, they require the use of facilitators and an investment of time by the other pertinent staff such as engineers and human resource professionals. Often companies use the former middle managers as facilitators.

The major long-term benefit of using teams though is their involvement of the line workers in the improvement process. The line worker knows more about what actually is occurring on the job than the managers can know. Teams provide the line worker a means of making improvements in the daily work processes. Another benefit of teams is that they increase the flexibility of the firm. When the environment of the firm requires it to respond quickly to customers, organizing the work into a team structure is an obvious response.

Teams of well-trained employees allow the firm to respond to changes faster because the members are flexible enough to respond to changes in the product requirements. Team members are also cross-trained to perform the different jobs on the team, which means that when a fellow team member falls behind on a job, others are able to help, so that the team will meet its production goals. This helps to eliminate variance in the production times. Teams also pay for themselves by increasing the efficient use of the team labor. Because the members are flexible and teams allow sharing of resources and the division of jobs, a worker who is temporarily prevented from working on one part of a job can move to another part of a job and help a co-worker there.

This ability of team workers to assist each other is greatly enhanced by locating the team's operations in a layout that facilitates sharing the work. Group technology layouts or cellular manufacturing layouts complement the use of teams. These types of layouts allow team workers fast access to different parts of the process, so there is little delay in switching from one task to another.

15 Materials Management

Materials have to be actively managed at each stage in the supply chain. From the view of an individual firm, materials must move into the firm, materials must be moved within the firm, and transformed materials must be moved outside of the firm to the customer. The supply chain moves materials to satisfy customers. In this chapter we examine the performance of an individual firm somewhere in the supply chain. How well that firm manages its own material movements determines how well it performs.

Each firm in the supply chain typically has three functions involved in its material management. These are purchasing, manufacturing (including production control), and distribution (including marketing). Each group needs to be effective and efficient to ensure that the needed materials arrive on time, are produced on time, and are shipped on time. As we discussed earlier, to ensure that the overall system is performing as well as possible, it is necessary that goals be set for the entire system and that there be system-wide performance measures that will induce each function to perform in a way that improves the entire system.

Little's Law

To understand the flow of materials through the system, we will examine the flow using Little's Law. Little's Law is a well-understood relationship between the average amount of inventory (I) in the system, the average throughput (T) of the system, and the average flow time (F) of the system. The relationship is that $I = T * F$. This means that the average amount of inventory in any system is equal to the average throughput multiplied by the average flow time. So, if on average a system is able to process 100 parts per hour (T = 100 parts/hour) and the average system flow time is 20 hours, then there will on average be 2,000 parts in inventory in the system.

This relationship can be examined in different ways. If we are interested in knowing how a proposed increase in throughput to 120 parts/hour without a change in flow time will affect inventory, we can estimate the effect on inventory by calculating I = 120 * 20 = 2,400 parts. This means that on average after the change we will have 2,400 parts in inventory.

But what is more valuable than estimating the effect that changes will have on the system is using Little's Law to understand the relationships between the functions involved in the material flow. If the flow time for orders through purchasing is 4 weeks (i.e., it takes 4 weeks from the time a product is ordered by operations until it is delivered to operations) and if purchasing's throughput is 100 orders a week, then there will be I = T * F orders in the system, or I = 100 * 4 = 400 orders in the system. These will be in the form of paperwork sitting in purchasing, orders being prepared by suppliers, orders in transit, and orders sitting on the dock in the receiving department.

Another example of using Little's Law is to examine the flow through the manufacturing or operations function. For operations, if the throughput of orders is 100/week and its throughput is 1 week for each order, then there will be inventory equal to 100 orders in the operations system (i.e., I = 100 orders/week * 1 week = 100 orders). If this company is a make-to-order company, then the orders will be shipped when they are finished and sent to shipping by operations.

If shipping attempts to reduce freight bills by holding orders to compile them into larger shipments, the order may sit in shipping for a while. If the average throughput of shipping is 100 orders a week and its average flow time is 1 week, then the inventory sitting in shipping is 100 orders (i.e., I = 100 orders/week * 1 week).

The relationships between the functions and how they affect inventory is illustrated in Figure 15.1, which also shows that we can test the impact of each department's operation on the entire system. The sum of the inventory in each department should sum to the inventory in the entire system and the relationship of Little's Law should hold for the entire system. This analysis can be seen in the top portion of Figure 15.1, where it shows that the system throughput (i.e., orders shipped out) is 100 orders/week and that the system flow time is 6 weeks (4 weeks in purchasing, 1 week in operations, and 1 week in shipping). Solving for I for the system gives the result that I = 600 orders, which is just the number of orders that is in the system as shown in the lower portion of Figure 15.1.

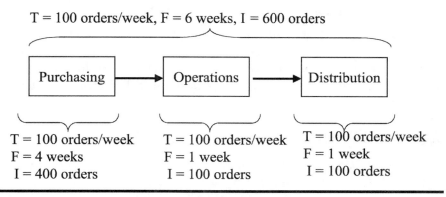

T = 100 orders/week, F = 6 weeks, I = 600 orders

Purchasing → Operations → Distribution

T = 100 orders/week
F = 4 weeks
I = 400 orders

T = 100 orders/week
F = 1 week
I = 100 orders

T = 100 orders/week
F = 1 week
I = 100 orders

Figure 15.1   Analyzing the System with Little's Law

Analyzing the system with Little's Law helps us identify many ways of improving this simple system. Can we change something in purchasing to reduce the amount of time it takes to get an order? Can we develop another method of reducing shipping costs, rather than holding the average order for a week before we ship it?

Inventory Management

From Little's Law as shown above, we can see that there is a relationship between the amount of inventory that we have in stock and the flow time (F). This relationship is F = I/T. So, if we have an inventory of 1,000 units and a throughput of 100 units/week, then the average time it takes for inventory to flow through our system is F = (1,000 units)/(100 units/week) = 10 weeks.

Inventory management—The branch of business management concerned with planning and controlling inventories.
APICS Dictionary, 9th edition, 1998

Although we discussed inventory in the earlier JIT section as being "the root of evil," inventory is necessary to satisfy the customer. The goal is to be able to hold the minimum amount of inventory necessary and still be able to provide the desired level of customer service. For this reason, inventory management is usually seen as an essential task of materials management.

Inventory contributes to value if it positively impacts one of the terms in the value equation. If inventory can reduce costs (such as it will do when buffer stock is put in front of a bottleneck), it increases value. If inventory allows faster response to a customer (who wants and is willing to pay for faster response) by having inventory available for the customer, it increases value. There are many other ways in which inventory can increase value, but once we determine the level of inventory we need, we must then create a system to manage that inventory.

Types of Inventory

There are two basic types of inventory. There is independent-demand inventory. This refers to inventory that is used to satisfy the demands of the end consumer. This inventory is a finished goods inventory for the supply chain. The second type of inventory is dependent-demand inventory. Dependent-demand inventory is used to fabricate parts or components for the final product that goes to the end consumer.

Each of these types of inventory behaves differently and has to be managed differently. The major distinction between them is how their demand is managed. The demand for the independent inventory must be forecast as explained in Chapter 8. The demand for the dependent inventory can be calculated, because the amount of the type of inventory that is needed is directly related to the demand for the final product. For example, the demand for some components, which are in a finished item, is always dependent on the number of a finished item that we need. First, we will examine how to manage independent inventory. Then we will investigate the methods to manage dependent inventory.

Designing the Inventory Management System for Independent Inventory

Inventory must be managed. An inventory system is a set of rules and procedures which are developed to manage a firm's inventory to ensure that the firm can meet its goals at the lowest possible cost to itself.

All inventory management systems answer two questions: How much to order, and when to order. The rules established to answer these two questions must be continuously reviewed and revised as appropriate. As demonstrated using Little's Law earlier in the chapter, the answer to these questions depends on other characteristics in the system. For example, as the flow time of an item becomes shorter, the system will need less inventory, so the two questions of *when* and *how much* need to be answered again.

Figure 15.2 shows the changes in the inventory level of an item as customers purchase it and as the item is restocked. The item in Figure 15.2 has a steady

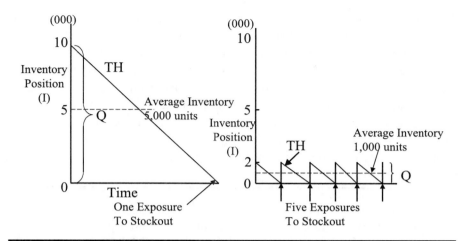

Figure 15.2   Order Size and Stockout Possibilities

demand, which is represented by the slanting line labeled TH for throughput. New material is received in order sizes of Q when the inventory position (I) is zero as shown in Figure 15.2. Depending on our decisions about how often an order is placed and how much we order, the graph of the inventory position over time will change. As shown in the left diagram of Figure 15.2, if we order only once a year, we have a larger average inventory, but there is only one time a year we will stock out. If we order 5 times a year as shown in the left diagram of Figure 15.2, we have a smaller level of inventory, but there are more opportunities of stocking out of the item, since every time our inventory position approaches zero there is a chance the new order will not be received on time.

Inventory costs money. A company must pay to purchase materials, it must pay to transform these materials into something else, and it must pay to store and move these materials around. Inventory costs are one input into answering the question of how much to order and when to order. Typically, companies estimate the cost of holding inventory, so they can understand how it influences their ability to make a profit. Usually the inventory holding costs consists of the opportunity cost of capital, the costs of taxes and insurance, the costs of handling and storing the inventory, and finally the costs of lost, stolen, damaged, or obsolete inventory. The accounting department estimates the costs for each period and reports this costs to all of the material managers.

The opportunity cost of the capital that is tied up in inventory can be significant. The opportunity cost is the cost of the return on investment the company could obtain if it was able to invest its money in something other than inventory. If a company has a 20% annual return on investment, then its costs of capital will be set at 20% per year. The other costs may be totaled up over a year's time and then the percent of these costs can be calculated as a percentage of the total cost of inventory. It may be that our firm had $1,000,000 in inventory last year and it cost $100,000 to hold this inventory. This is a cost of 10% a year. The combined holding cost is 30%/year (20% + 10%).

The holding cost is calculated as H = (unit price)*(holding cost rate). Given the above figures, the annual holding cost of an item which cost $100 to purchase or to manufacture would be $30 a year (i.e., $100 * 30%).

Carrying cost—The cost of holding inventory, usually defined as a percentage of the dollar value of inventory per unit of time (generally one year). Carrying cost depends mainly on the cost of capital invested as well as such costs of maintaining the inventory as taxes and insurance, obsolescence, spoilage, and space occupied. Such costs vary from 10% to 35% annually, depending on type of industry. Carrying cost is ultimately a policy variable reflecting the opportunity cost of alternative uses for funds invested in inventory.

APICS Dictionary, 9th edition, 1998

Opportunity cost—1) The return on capital that could have resulted had the capital been used for some purpose other than its present use. 2) The rate of return investors must earn to continue to supply capital to a firm.

APICS Dictionary, 9th edition, 1998

Ordering cost—Used in calculating order quantities, the costs that increase as the number of orders placed increases. It includes costs related to the clerical work of preparing, releasing, monitoring, and receiving orders, the physical handling of goods, inspections, and setup costs, as applicable.

APICS Dictionary, 9th edition, 1998

Another cost associated with inventory is the cost of ordering or of setting up the machines to produce the inventory. This cost exists when there is a fixed cost to make an order or to set up a machine. For example, it may take 2 or 3 setup technicians 30 minutes to set up a machine. If those technicians could be doing something else during that time, then the setup cost would be the cost of 30 minutes of their time. Order costs may include the cost of a purchasing agent investigating and placing the order as well as a share of the fixed costs of operating the purchasing department.

The equation for total inventory cost is then the holding cost plus the carrying costs, or:

$$TC = (D_a/Q)S + (Q/2)H$$

The first term in this equation [i.e., $(D_a/Q)S$] represents the annual demand for an item (D_a) divided by the order quantity (Q). The result of this division is the number of times an order is placed for an item each year. Since the cost of placing this order is S, the first term represents the cost of placing orders for an item over the period of a year. The second term in the equation $(Q/2)H$ says that the average inventory $(Q/2)$ multiplied by the cost of holding a unit in inventory (H) is the expected annual inventory carrying cost for a particular item.

For example, if a shovel costs a hardware store $30 to purchase, the holding cost rate is 20%, the annual demand is 2,000 shovels, the order quantity is 100, and the order cost is $100, then we can calculate the total cost of inventory as:

$$TC = (2,000/100)\$100 + (100/2)(\$30 * 20\%) = \$2,300.$$

Inventory Policy

Given the above background, we can look at inventory policies to answer the question of when to order. This inventory policy tells us when the level of inventory or the inventory position of an item should be reviewed to identify whether an item should be ordered. Until the computer took over most inventory functions, the inventory policy usually used was a periodic review policy. With this policy the inventory manager would count the amount of inventory available at prescribed times. Now that computers can continuously review the inventory status of thousands of item, most firms use a continuous review policy. This means that every time an item is used, the computer calculates the balance in stock (i.e., determines the item's inventory position) and evaluates whether this inventory position is at or below a reorder point.

Fixed reorder cycle inventory model—A form of independent demand management model in which an order is placed every n time units. The order quantity is variable and essentially replaces the items consumed during the current time period. Let M be the maximum inventory desired at any time, and let x be the quantity on hand at the time the order is placed. Then in the simplest model, the order quantity will be $M - x$. The quantity M must be large enough to cover the maximum expected demand during the lead time plus a review interval. The order quantity model becomes more complicated whenever the replenishment lead time exceeds the review interval, because outstanding orders then have to be factored into the equation. These reorder systems are sometimes called fixed-interval order systems, order level systems, or period review systems.

APICS Dictionary, 9th edition, 1998

Fixed reorder quantity inventory model—A form of independent demand item management model in which an order for a fixed quantity, Q, is placed whenever stock on hand plus on order reaches a predetermined reorder level, R. The fixed order quantity Q may be determined by the economic order quantity, by a fixed order quantity (such as a carton or a truckload), or by another model yielding a fixed result. The reorder point R, may be deterministic or stochastic, and in either instance is large enough to cover the maximum expected demand during the replenishment lead time. Fixed reorder quantity models assume the existence of some form of a perpetual inventory record or some form of physical tracking, e.g., a two-bin system, that is able to determine when the reorder point is reached. These reorder systems are sometimes called fixed order quantity systems, lot-size systems, or order point-order quantity systems.

APICS Dictionary, 9th edition, 1998

Either type of inventory model will calculate a reorder point, which is the answer to the question of when to order. When calculating the reorder point, there are four factors to consider. The first is the length of time needed to replenish the order. The second is the average demand during the order replenishment lead time. And, the third is the variance in the demand pattern and the delivery time. Another piece of information is needed. That is, what is the level of customer demand that will be satisfied? Do we want to carry enough inventory to meet the needs of 99% of our demand or just 80% of the demand? This is a strategic decision that requires knowledge of the business and customer expectation to answer correctly.

Demand for an item is seldom level. It has randomness, so we usually speak of the average demand. But, if we carry only enough inventory to meet the average demand, we will be out of stock 50% of the time before we receive our new shipment of inventory. This is illustrated in Figure 15.3, where the average demand under the normal curve is at $Z = 0$. As we move to the right of this point we have additional inventory to meet demand, when it is greater

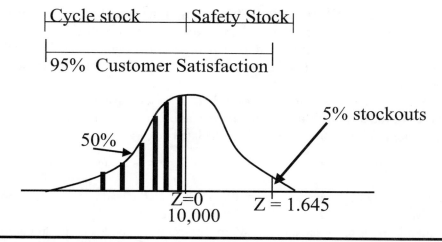

Figure 15.3   Setting the Reorder Point

than the average. The question becomes how much should we carry. In Figure 15.3 we see that if we carry enough inventory to meet 95% of the demand, then we are at point $Z = 1.645$. In technical terms, this means that we are 1.645 standard deviations above the mean. All of the inventory that we carry in addition to that inventory needed to meet average demand is called *safety stock*. The inventory that we need to meet average demand is called *cycle stock*.

Cycle stock—One of the two main conceptual components of any item in inventory, the cycle stock is the most active component, i.e., that which depletes gradually as customer orders are received and is replenished cyclically when supplier orders are received. The other conceptual component of the item inventory is the safety stock, which is a cushion of protection against uncertainty in the demand or in the replenishment lead time.

APICS Dictionary, 9th edition, 1998

The reorder point is the average demand during lead time (D_{LT}) plus safety stock (SS) that is needed because of either variance in demand during lead time or variance in the delivery time. Again, the average demand during lead time is known as the cycle inventory. This is illustrated in Figure 15.3 using the normal curve. In Figure 15.3, the cycle inventory is the inventory needed to meet the average demand, which means that 50% of the time there will be enough or more than enough inventory to meet demand. But, if we want to have a service level greater than 50%, we need to have more inventory. All inventory greater than that needed to serve the average demand is safety stock inventory. This is shown as the space between the mean and the line towards the upper right tail of the curve.

For example, if demand for shovels is 40 a week on average, with a delivery lead time of 1 week and a standard deviation of 5 shovels a week, the reorder point for a 95% service level is calculated below:

$$ROP = D * LT + z * \sigma = 40 * 1 + 1.645 * 5 = 48.225 \text{ shovels}$$

In this equation, the average demand is given as D, the lead time for demand is given as LT, z is the number of standard deviations above the average demand that is to be satisfied (1.645 in this example), and σ is the standard deviation during lead time (i.e., 5).

The second question to be answered by the inventory model is how much to order. This question is answered by the order quantity rule in use. The economic order quantity rule is explained in the next section. But note, that in many situations, it is suitable to order just what has been used. When this is the case, then the rule that is in use is the Lot-for-Lot (LFL or L4L) order rule.

Order Quantity Rules

Order quantity rules were created to minimize the total costs of inventory. They can be used with both independent- and dependent-demand inventories. There are a wide variety of order quantity rules, which consider many different costs and situations. One of the simplest rules and the best known is the economic order quantity (EOQ). This rule minimizes the costs of holding inventory and the costs of ordering inventory. It is derived from the inventory total cost equation given in Equation 1.

$$EOQ = \sqrt{\frac{2DS}{H}} \qquad (1)$$

In Equation 1, D is the annual demand for a given part or product, while S is the setup cost or order cost for the part and H is the holding cost for the part. H is found by multiplying the price of the part by the holding cost rate. For example, if the annual demand for a shovel is 2,000 shovels with an acquisition cost of $30, a holding cost of 20%, and an order cost of $100, the EOQ would be calculated as shown in Equation 2.

$$EOQ = \sqrt{\frac{2 * 2,000 * \$100}{20\% * \$30}} = 258.2 \text{ shovels} \qquad (2)$$

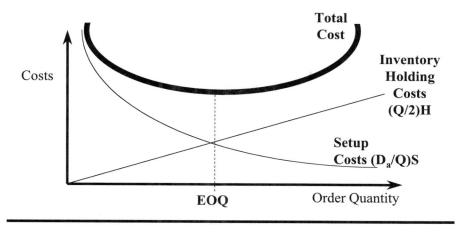

Figure 15.4   Economic Order Quantity (EOQ)

To simplify the discussion of the EOQ, it is illustrated in Figure 15.4. One reason that this heuristic has remained popular with materials managers for decades is that the total cost curve is flat around the EOQ point on the *x* axis. This means that even if we are not able to estimate the costs of holding and ordering inventory correctly, we still have a good chance of identifying a point which is close to the lowest cost quantity, because our calculations will give us an order quantity which is on the flat part of the curve.

There are some basic assumptions that the EOQ model makes that may not be true in any given situation. The EOQ model assumes that demand for an item is constant over time. It also assumes that there are no order discounts, so that the price of the item is fixed. And a final important assumption is that the EOQ rule assumes that delivery lead time is known and is constant.

Inventory Classification

One of the simplest methods of managing inventory is to classify the inventory according to its importance. The inventory's importance is usually calculated as the amount of the inventory that flows through the system and the cost of that inventory. The most important items to manage are those which have a high dollar value and a high volume through a facility. A common way of doing this is called ABC Analysis and is illustrated in Figure 15.5.

The goal of doing this analysis is to have management focus its attention on the most important types of inventory in the system. This analysis helps

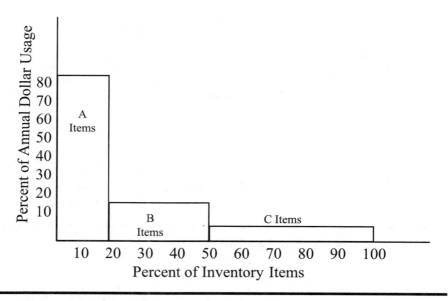

Figure 15.5   Graph of ABC Analysis

everyone recognize which types of inventory are the most important. For example, if you were managing a hardware store, there would be limited value in spending much time managing the inventory of bulk nails, which you sell by weight, since their annual sales multiplied by their low unit costs would show that the amount of cash flow related to nails was not significant compared with the cash flow connected with power tools. This type of analysis would suggest to the hardware store manager that it is more important to manage the inventory of the power tools very carefully.

The ABC analysis in Figure 15.5 is done by listing the amount of each item in stock sold during the last year. The quantity sold is multiplied by the purchase price of each item. This product is then used to rank each of the items in descending order. So, the item at the top is the item with the largest dollar volume and the item at the bottom has the lowest dollar volume. We can then classify items in one of two ways. The easiest is to classify them according to the x axis in Figure 15.5. That is, we count the number of items and state that the first 20% of the items with the highest dollar volume are A items. The next 30% of items are B items and the final 50% of the items are C items. The other way to classify the items is using the y axis. Using this approach, we have to calculate the total dollar volume of inventory annually and then calculate the percent of this total for which each item in inventory accounts. Starting at the top, we classify inventory as A items until the cumulative percentage in A is equal to 80%. Continuing down the list, we classify inventory as B items until another 15% of the total dollar volume of inventory has been included. This leaves only 5% of the total dollar volume of inventory, which is classified as C items. Regardless of the approach that we take, we will obtain almost exactly the same listing for each class of inventory.

A typical example of a C item would be a consumable supply. Consumable supplies can often be stocked at higher than necessary levels since they are cheap and we do not want to cause delays in any of our processes if they are not available.

Cycle Counting

A major function of management is control. Cycle counting is a technique that is used to control inventory. One of the most common failures in inventory systems is that the data are not accurate. This error can be prevented. One common technique is to bar code inventory, so that the amount of inventory and the type of inventory at each stage can be quickly scanned into the computer. But, it is also important that cycle counts be conducted.

Cycle counting is a method of counting all of the items in inventory on a rotating basis. The frequency with which a particular type of inventory is counted can reflect its classification in the ABC analysis discussed earlier, but all inventory needs to be counted on some regular basis. For example, the inventory could be divided into 20 sets. Each set would then be counted once in every 20 days. This physical count of actual inventory and the reconciliation of the computer records will help to improve accuracy. In addition, the cycle count will provide information about where bad data are entering the system, or where data are not entering the system.

Cycle counting—An inventory accuracy audit technique where inventory is counted on a cyclic schedule rather than once a year. A cycle inventory count is usually taken on a regular, defined basis (often more frequently for high-value or fast-moving items and less frequently for low-value or slow-moving items). Most effective cycle counting systems require the counting of a certain number of items every workday with each item counted at a prescribed frequency. The key purpose of cycle counting is to identify items in error, thus triggering research, identification, and elimination of the cause of the errors. One approach to reducing the inventory to what is needed is similar to the two bin method of inventory control described earlier. The average demand is first estimated and a moderate level of safety stock is determined. Inventory in excess of this level is removed and placed into a separate bin. The bin with the additional inventory is available, but cannot be readily accessed. In this way inventory usage can be computed and over a period of weeks or months, a safety stock level high enough to ensure good service can be determined. While that is going on, the extra bin is used to restock the consumable supplies closet, but the remote bin is not itself restocked, so the amount of inventory in the system is reduced. When the remote bin is empty, there should be enough data to know how much inventory to carry and there should be less inventory in the system.

APICS Dictionary, 9th edition, 1998

Materials Requirements Planning

When demand is dependent, the amount of material needed can be calculated and not forecasted. The end items produced by the factory (i.e., those scheduled in the master production schedule) are the only independent-demand items in the factory. The components needed to produce those end items are dependent on the demand for the end items. It is the task of materials requirements planning to calculate how many of each component are needed and when they are needed.

There is some confusion about Materials Requirements Planning (MRP), because much of the software that has been developed to do materials requirements planning uses the same name. But, the task of materials requirements planning must be done regardless of whether there is software to do it.

Material requirements planning (MRP)—A set of techniques that uses bill of material data, inventory data, and the master production schedule to calculate requirements for materials. It makes recommendations to release replenishment orders for material. Further, because it is time-phased, it makes recommendations to reschedule open orders when due dates and need dates are not in phase. Time-phased MRP begins with the items listed on the MPS and determines (1) the quantity of all components and materials required to fabricate those items and (2) the date that the components and material are required. Time-phased MRP is accomplished by exploding the bill of material, adjusting for inventory quantities on hand or on order, and offsetting the net requirements by the appropriate lead times.

APICS Dictionary, 9th edition, 1998

Whether it is done by hand or using computer software, all MRP systems use the same logic to calculate when items needed to make another item must arrive at the factory or warehouse to be available. The materials requirements planning stage considers both purchased materials and manufactured components. The MRP logic will be explained in the following sections. First, the inputs to the system will be explained. Then the mechanics of the MRP planning process will be explained, and finally the outputs of the MRP planning process will be explained.

MRP Inputs

The inputs to the MRP system are the master production schedule (MPS), the bill of materials (BOM), and an inventory records file, which has both the quantity of inventory in stock and the lead times of each item.

MRP requires that each end item have a BOM, which clearly identifies the components needed to make the end item. As explained in Chapter 9, the BOM can be represented in two different ways. One method is to show it as a hierarchical table as shown in Table 15.1. The other is to show it as a product structure tree as in Figure 15.6. Both of these represent the same BOM. But, note that the table has more information than the product structure tree does; the table includes the lead time needed to either manufacture or purchase the item within the BOM.

Table 15.1   BOM for End Item A

Level	Item No.	Description	Number Required	Lead Time
0	A	End item		1 week
1	B	Component for assembly	2	1 week
2	D	Purchased subcomponent for B	1	1 week
2	E	Subcomponent for B	1	1 week
3	F	Purchased subcomponent for E	3	2 weeks
1	C	Purchased component for assembly	1	1 week

The MRP calculation also requires that the MPS be known. It is generally recommended that the planning horizon for the MPS be at least as large as the average lead time of the end items produced by the facility. This allows the MPS to be stable, which then means that the requirements being conducted in the MRP have some validity.

The inventory records for all the end items and for the components of these end items need to be available. Needless to say, they need to be accurate. Otherwise, creating an MRP will be an exercise in futility. Garbage in gives garbage out. When computerized MRP failures are discussed, the lack of

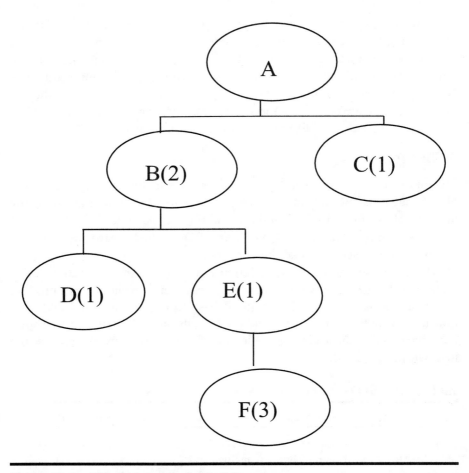

Figure 15.6   Bill of Materials for End Item A

accuracy of the inventory records is usually one of the major reasons a system failed.

MRP Logic

MRP planning starts with the end item as given in the MPS, and determines when that item must be started to be completed on time. That date is then used as the order due date for the next level in the BOM. The logic works backwards until it reaches the end of the BOM. This is illustrated with a Gantt chart in Figure 15.7.

The Gantt chart shows that before one step in the process can begin, other steps need to done completely. For example, before A can begin, both B and C have to be completed and waiting in the shop. The Gantt chart also shows that

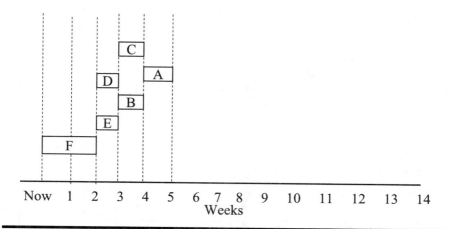

Figure 15.7   MRP Gantt Chart

for A to be completed on time, F has to be in stock and ready now, because there is no time left in the schedule for delays.

Requirements explosion—The process of calculating the demand for the components of a parent item by multiplying the parent item requirements by the component usage quantity specified in the bill of material.

APICS Dictionary, 9th edition, 1998

Item Master	Part A - End item	Lead Time 1 wk Safety Stock								
	Order Q. LFL									
	Scrap									
Inventory Status		1	2	3	4	5	6	7	8	
	Gross Requirements					100				
	Scheduled Receipts									
	Projected On Hand 0	0								
	Net Requirements					100				
	Planned Order Receipt					100				
	Planned Order Release				100					

Figure 15.8   Materials Requirements Plan for Item A

Part A: Lead Time = 1 week OQ = LFL						Safety Stock = 0			
	1	2	3	4	5	6	7	8	
1. Gross Requirements					100				
2. Scheduled Receipts									
3. On Hand 0	0	0	0	0	0				
4. Net Requirements					100				
5. Planned Receipt					100				
6. Planned Order Release				100					
Part B: Lead-time = 1 week OQ = LFL						Safety Stock = 0			
1. Gross Requirements				200					
2. Scheduled Receipts									
3. On Hand 0	0	0	0	0					
4. Net Requirements				200					
5. Planned Receipt				200					
6. Planned Order Release			200						
Part C: Lead-time = 1 week OQ = LFL						Safety Stock = 0			
1. Gross Requirements					100				
2. Scheduled Receipts									
3. On Hand 0	0	0	0	0					
4. Net Requirements					100				
5. Planned Receipt					100				
6. Planned Order Release			100						
Part D: Lead-time = 1 week OQ = LFL						Safety Stock = 0			
1. Gross Requirements			200						
2. Scheduled Receipts									
3. On Hand 100	100	100	0						
4. Net Requirements			100						
5. Planned Receipt			100						
6. Planned Order Release		100							
Part E: Lead-time = 1 week OQ = LFL						Safety Stock = 0			
1. Gross Requirements			200						
2. Scheduled Receipts			100						
3. On Hand 0	0	0	0						
4. Net Requirements			100						
5. Planned Receipt			100						
6. Planned Order Release		100							
Part F: Lead-time = 2 weeks OQ = LFL						Safety Stock = 0			
1. Gross Requirements		300							
2. Scheduled Receipts		100							
3. On Hand 200	200	0							
4. Net Requirements		0							
5. Planned Receipt									
6. Planned Order Release									

Figure 15.9   MRP Explosion of Item A

The construction of the materials requirements plan is illustrated in Figure 15.8 for end item A. Note that 100 of item A are required in period 5 and there is none on hand right now and none is scheduled to be received, so the Projected On Hand in period 1 is 0. Since there is no Gross Requirements for A until period 5, the Projected On Hand stays at 0 until then, when it becomes a negative 100 or (100). To have enough A in period 5, we need to have a Planned Order Receipt for A in period 5. Since item A has a 1 week lead time this means that we need to have a Planned Order Release in period 4. So, we schedule 100 to be released in period 4.

As shown in Figure 15.9, this MRP explosion continues down through the BOM (see Table 15.1) until the releases and quantity needed for each item in the BOM are calculated. It is very important to note that the MRP explosion is hierarchical. This means that everything on level 1 is calculated before anything is calculated for level 2, etc. So, in Figure 15.9 the gross requirements for B and C are calculated before the requirements for lower level items such as D. Since 2 B are needed for each A, there is a gross requirement for 200 B in period 4. Following the logic described for A, this results in a gross requirement for 200 B in period 3. In a similar manner we calculate the gross requirements for C in period 4, which is 100, since we only need 1 C for each A. The gross requirement for D is based on the planned order release of its parent B. So, the gross requirement for D in period 3 is 200. Since there are 100 D in stock, the planned order release for D is just 100 in period 2. In a similar manner we calculate the gross requirements for item E, based on the planned order releases of its parent B. E had a scheduled receipt of 100 in period 2, which was subtracted from the gross requirements of 200 in period 3. This combined with the planned receipt of 100 in period 3 combined to give a planned on-hand of 0 items of E in inventory in period 3. As shown in the BOM, E is the parent of F, so E's planned order release becomes the gross requirements of F. So, F has a gross requirement of 300 in period 2, since 3 F are required for every E. Since 200 F are on hand and 100 more will be received in period 2, there is no net requirement for F in period 2.

MRP Outputs

All of the calculations above are completed for all end items in the MPS. These calculations tell the planner when to release orders for production or when to release purchase orders to suppliers. They also tell the shop the due dates of the work that is going to each machine.

As can be seen from the example above, it is very tedious doing these calculations by hand. So, unless the BOM is very flat or there are very few items in the BOM, MRP software is used to perform the planning.

The MRP software has some advantages in that it can send messages to a firm's suppliers and it can send the same information to everyone in the firm. The distribution department will know when to expect items into finished goods and purchasing will know what items will need to be purchased.

In today's environment many MRP systems are only one module in a larger system such as an Enterprise Resource Planning (ERP) system. The ERP system allows the MRP data to be combined with other company data so that the firm can project its cash flow, monitor its inventory and delivery performance, etc. ERP systems will be discussed in Chapter 16.

Capacity Requirements Planning

One problem with the MRP logic is that it ignores the load on each work center when it was determining the schedules. This means that one work center could be a bottleneck and might not be able to complete its portion of the work on time. Since this process is a series of dependent events, that would inevitably lead to disruption throughout the shop. This in turn would lead to higher inventory levels, longer flow times, and missed shipments. To avoid these problems, companies examine the capacity requirements and make changes as necessary.

Capacity requirements planning (CRP)—The function of establishing, measuring, and adjusting limits or levels of capacity. The term capacity requirements planning in this context refers to the process of determining in detail the amount of labor and machine resources required to accomplish the tasks of production. Open shop orders and planned orders in the MRP system are input to CRP, which through the use of parts routings and time standards translates these orders into hours of work by work center by time period. Even though rough-cut capacity planning may indicate that sufficient capacity exists to execute the MPS. CRP may show that capacity is insufficient during specific time periods.

APICS Dictionary, 9th edition, 1998

Many MRP systems provide load reports similar to those explained in Chapter 9 for each work center by period. It is then possible to tell if any work center is overloaded in any given period. When an overloaded work center is identified, the planner then has to remove some of the work. That can be done by changing the MPS so that some items are needed earlier or some can be completed later. For the schedule to be feasible, the loads should be within the capacity of each work center, or they will not be completed on time and the entire schedule will fail.

Load—The amount of planned work scheduled for and actual work released to a facility, work center, or operation for a specific span of time. Usually expressed in terms of standard hours of work or, when items consume similar resources at the same rate, units of production.

APICS Dictionary, 9th edition, 1998

---

16 Purchasing and Distribution

I n many firms purchasing has been seen as a clerical activity. However, the emergence of the supply chain management concept has enlightened many managers about the strategic role played by purchasing.

Purchasing contributes to the firm's efficiency and effectiveness in many ways. First, it helps to determine a firm's cost structure through negotiations with suppliers. Reducing the investment in inventory and improving the quality of incoming parts and components through its vendor selection and supplier development policies. Encouraging new product development by encouraging supplier involvement in technology development.

Purchasing—The term used in industry and management to denote the function of and the responsibility for procuring materials, supplies, and services.

APICS Dictionary, 9th edition, 1998

Savings obtained by the purchasing function go immediately to the firm's bottom line. If purchasing can reduce the amount paid to vendors by $1,000,000 during a year, the firm has $1,000,000 more in profit. If a firm obtains $1,000,000 more in sales, and it has an after tax profit margin of 10%, then profits are increased by only $100,000.

An important role performed by purchasing is boundary spanning. Purchasing is in position to obtain information not only about prices and availability of goods, but also about new supply sources and new technology in the market.

Purchasing Cycle

A typical purchase follows a similar set of steps in almost every organization. First, there has to be a recognition of a need for a purchase to be made. This need may be identified through the use of an MRP system, a forecast, or a stockroom cycle count. To communicate this need to purchasing, whatever function determined there was a need for an item fills in some type of order form, for example, a purchase requisition form. Purchasing then determines the potential sources for obtaining the material. If it is a routine purchase, there will ordinarily be suppliers' names and information on file in the purchasing office. If it is not a routine purchase, the purchasing agent will search through directories, obtain recommendations from users, search the internet, or request information from potential suppliers.

Purchasing agent—A person authorized by the company to purchase goods and services for the company.
APICS Dictionary, 8th edition, 1998

Once sources have been identified, purchasing typically invites the sources to bid by sending a request for proposal (RFP) or quotation to them. In the age of the Internet, some potential suppliers have been able to automate this by providing web pages that allow a potential customer to submit their RFP to them electronically. If the request is for an item similar to what has been manufactured by the company, it is sometimes possible for the purchasing department to quickly receive an electronic bid.

Request for proposal (RFP)—A document that describes requirements for a system or product and requests proposals from suppliers.
APICS Dictionary, 8th edition, 1998

Once the purchasing department has a bid that is acceptable (which means that the provider is reputable, the product to be purchased meets specifications, and the price is appropriate), the purchasing department prepares a purchase order (PO). The formality of this PO may depend on the size of the purchase. If it is an expensive one-time buy, the purchase order may be very detailed. If it is a low-cost item, such as the purchase of a training manual or book, it may be done over the phone or by fax.

Purchase order—The purchaser's authorization used to formalize a purchase transaction with a supplier. A purchase order, when given to a supplier, should contain statements of the name, part number, quantity, description, and price of the goods or services ordered; agreed-to terms as to payment, discounts, date of performance, and transportation; and all other agreements pertinent to the purchase and its execution by the supplier.

APICS Dictionary, 8th edition, 1998

Often when the items being purchased are used by production, purchasing will monitor the progress of the order. It will evaluate progress by the supplier and keep operations informed about any problems. This may be done with some type of purchasing follow-up report or via information that is posted on the company's intranet.

Firms have a receiving function to ensure that the material and equipment they actually receive are what the company ordered and paid for. Depending on the firm's relationship with its supplier, receipt of the material may be confirmed by the supplier itself, or it may be received by a separate function which performs only this task. The receiving function checks the quantity of material received and may perform quality control checks on it or send it to a quality control lab to do this. The receiving function records discrepancies and logs the receipt of the material into the system and then places it in its appropriate storage location until it is needed.

Receiving—The function encompassing the physical receipt of material, the inspection of the shipment for conformance with the purchase order (quantity and damage), the identification and delivery to destination, and the preparation of receiving reports.

APICS Dictionary, 8th edition, 1998

Receiving report—A document used by the receiving function of a company to inform others of the receipt of goods purchased.

APICS Dictionary, 8th edition, 1998

Once the material is received by the firm, it has to be paid for. This is usually done by accounting. But, accounting needs to reconcile the invoice received from the vendor with the receiving report and the supplier needs to resolve any discrepancies.

Purchasing performs another function during this purchasing cycle. It maintains records about suppliers and their performance as well as records about the receipt of materials and the sources of these materials.

Supply Problems

Purchasing is involved not only in obtaining and receiving supplies into the firm, but also in resolving any problems with the supplies during their use. This requires purchasing to stay in communication with the users of the materials and supplies.

Typical supply problems that need to be resolved are a supplier's inability to meet delivery dates, nonconformance of materials to specifications, receipt of damaged materials, or receipt of the wrong quantity of materials. An additional problem is when there is a need to change the specification for the materials to be purchased. To resolve these quickly, purchasing must stay in communication with the suppliers of the materials and have clear procedures in place for resolving problems.

Logistics

Logistical activities include locating facilities, transporting material, storing inventory, communicating, and the handling associated with these. These activities have been integrated over the past 50 years and are an essential function of supply chain management. As distribution channels and the variety of products in them have increased over the past decades, logistics has assumed more importance in the supply chain.

Logistics—1) In an industrial context, the art and science of obtaining, producing, and distributing material and product in the proper place and in proper quantities. 2) In a military sense (where it has greater usage), its meaning can also include the movement of personnel.

APICS Dictionary, 9th edition, 1998

To achieve the highest level of service at the lowest possible cost, it is necessary for managers to examine the entire logistics system and not just one isolated facility or activity such as transportation. This permits the manager to conduct a total cost analysis of the entire system. The total cost analysis helps firms avoid optimizing the performance of one function such as warehousing while increasing costs even more at another function such as transportation.

Logistics system—The planning and coordination of the physical movement aspects of a firm's operations such that a flow of raw materials, parts, and finished goods is achieved in a manner that minimizes total costs for the levels of service desired.

APICS Dictionary, 9th edition, 1998

Total cost concept—In logistics, the idea that all logistical decisions that provide equal service levels should favor the option that minimizes the total of all logistical costs and not be used on cost reductions in one area alone, such as lower transportation charges.

APICS Dictionary, 9th edition, 1998

The analysis of the entire logistics system integrates all of the material in the previous chapters. The logistics system is concerned not only with the physical placement of the facilities, but also with the levels of inventory and the flow of material through those facilities. Each of these will be considered in the remaining sections of this chapter.

Warehouses

There are two major techniques to positioning a warehouse. These are to position it either to serve a particular market, or to serve a production plant. A market-positioned warehouse may serve as either a distribution center to other warehouses or provide goods directly to the retail store or the final customer. A market-positioned warehouse reduces costs by consolidating demand of many retail stores or customers, so that large quantities can be received from the manufacturer in one shipment. This reduces costs for the manufacturer and it reduces transportation costs, because a large volume of product can be shipped cheaply for the longer distance and then the higher cost local delivery can be made over a shorter distance. The production-positioned warehouse collects products manufactured at different plants.

This allows different products manufactured at different plants to be shipped together to a customer in the mix desired by the customer. The goal is to reduce transportation costs by allowing carloads of mixed product to move to the customer at the lower consolidated transportation rates.

Distribution center—A warehouse with finished goods and/or service items. A company, for example, might have a manufacturing facility in Philadelphia and distribution centers in Atlanta, Dallas, Los Angeles, San Francisco, and Chicago. *Distribution center* is synonymous with the term *branch warehouse*, although the former has become more commonly used recently. When a warehouse serves a group of satellite warehouses, it is usually called a *regional distribution center*.

APICS Dictionary, 9th edition, 1998

In some distribution channels, there might be intermediately positioned warehouses. These warehouses are located between the customer and the manufacturing plants. The purpose of these warehouses is to increase the level of customer service. They are sometimes necessary if a customer wants multiple products delivered at one time and they are produced long distances from each other.

Distribution channel—The distribution route, from raw materials through consumption, along which products travel.

APICS Dictionary, 9th edition, 1998

Intermediately positioned warehouse—A warehouse located between customers and manufacturing plants to provide increased customer service and reduced distribution cost.

APICS Dictionary, 9th edition, 1998

Total Cost Analysis of Logistics

To analyze the total cost of a logistics system, it is necessary to understand what activities each portion of the system is responsible for performing. For example, to what locations will a warehouse distribute product? Once we have identified that, we can examine various options and compare the total costs of

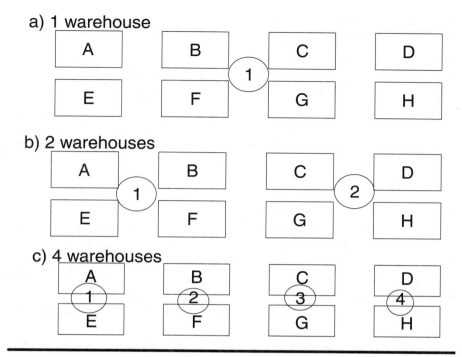

Figure 16.1   Total Costs of Logistics

each option. Lambert, Stock, and Ellram state that total costs are the sum of transportation costs, warehousing costs, order processing or information costs, lot quantity costs, and inventory carrying costs. To illustrate the use of total cost analysis, a simpler system involving just the total cost of the inventory in the system and the total transportation costs will be examined below.

Figure 16.1 has three parts, which show three different sets of warehouses to serve the same eight market areas. In Figure 16.1 a, there is one warehouse to serve all eight market areas. To simplify the analysis, it is assumed that each market has the same average weekly demand of 1,000 units with a standard deviation of 100 per week. This means that each market area has an annual demand of 52,000 units. If the warehouse is adjacent to a market area, it is assumed that the products can be delivered at the same cost throughout the market areas. If the warehouse is not adjacent to the market location, then the delivery charges are more expensive per hundredweight. The costs are shown in Table 16.1.

In this simple system, the number of warehouses influences both the average level of inventory in the system and the shipping costs. With one warehouse for the system, the standard deviation of demand on the warehouse

Table 16.1 Total Cost Analysis Example

Number Warehouses	Standard Deviation per Week	Warehouse Weekly Demand	ROP if LTL	ROP @TL	Average Inventory	Local Delivery	Total Cost
1	283	8,000	8363	15,363	7,823	4	676,454
2	200	4,000	4256	15,256	15,456	8	725,128
4	141	2,000	2181	15,181	30,645	8	1,028,909

from the eight markets is the square root of $(8*100^2)$. Remember, that the standard deviations cannot be added together, but variances can be added together, so the variances of demand from each market are summed and the square root of this sum is taken to find the standard deviation of demand on the warehouse. The weekly demand on each warehouse is the sum of the weekly demands from each market. When one warehouse serves all eight markets, the weekly demand on the warehouse is $8*1,000$ units per week or 8,000 units a week.

The service level desired in this system is 90%, which means that safety stock adequate to meet 1.282 standard deviations of demand on the warehouse must be carried at each warehouse. For example, if there is only one warehouse with a standard deviation of demand at the warehouse of 283, then $283*1.282 = 363$ units. The delivery time to the warehouse is 1 week. So, the required safety stock is added to the demand during the lead time to determine the reorder point (ROP) for each warehouse. For example, when there is one warehouse, the ROP becomes $8,000 + 363$ units or 8,363 units. But, this is less than 1 truck load (LTL), which means the warehouse would have to pay higher transportation costs.

To achieve the lowest transportation costs to the warehouse, the decision was made in this example to order only in truck load quantities (TL). In this example, that means that 15,000 units are ordered at one time. Since delivery time to the warehouse is one week, this means that the safety stock is still 363 if there is one warehouse, so the ROP is 15,363 units. The average inventory in the warehouse is the beginning inventory plus the ending inventory divided by two. If there is one warehouse, this is $(15,363 + 363)/2 = 7,823$ units. The ending inventory is 363 because this is the level of the safety stock and it is equally probable that the inventory will be higher than this as it is that the inventory will be lower than this. Regardless of the number of warehouses in the system, there will be a total of 28 deliveries to the warehouses, when the decision is made that all deliveries will be 15,000 units. A warehouse makes either a local delivery or a long-distance delivery each week to each market. So the total deliveries each week is 8. This means that with only one warehouse there will be 4 local deliveries and 4 long-distance deliveries. It costs $0.50/hundredweight for a local delivery and $1.00/hundredweight for a long-distance delivery. Each unit weighs 1 pound.

For the system in Figure 16.1a with 1 warehouse, there are then 4 local deliveries and 4 long-distance deliveries each week. The total cost of this system is the sum of the average inventory costs ($20/unit * average number of units) plus the cost of delivery to the warehouse and local delivery and long-distance delivery costs to the markets. For the one warehouse system this is $676,454. The total costs can be calculated for the other combinations of warehouses and markets served as shown in Table 16.1.

In the example given in Table 16.1, the system with 1 warehouse gives the lowest total costs. This is in part due to the low costs of local deliveries. If these charges were much higher than the costs of long-distance delivery, it might be cheaper to have 8 warehouses instead of 1 warehouse.

Integrated Logistics

Over the past decades as recognition of the advantages of integrating logistics functions has increased, there has been a trend toward organizing the functions under one executive. So, that one executive controls transportation, warehousing, inventory systems, order processing, purchasing, and production planning. The purpose of this centralized organization it to eliminate inefficiencies.

As the logistics functions become more integrated, they are able to achieve many efficiencies. But, a barrier to fully implementing an integrated logistics function is the lack of accurate information about costs. To effectively manage the trade offs in the logistics functions, it is necessary to understand the impact of each decision on the firm's cost. To do this, the logistics manager conducts a total cost analysis. But, many firms do not have accounting systems which can generate accurate cost data in a timely manner.

The logistics function requires more than accounting data that aggregates costs to report just the financial outcomes of decisions to the firm's stockholders. These reports usually work by allocating overhead costs to different products based on some criteria such as labor content. To do total cost analysis, the accounting records have to identify the costs associated with individual products, individual customers, and individual departments in the firm.

17 Information Technology and Supply Chain Management

S upply chain management is driven by the customer. It requires communication to all participants in the supply chain of the customer's needs and wants as well as how well these needs and wants are being met. To facilitate managing the linkages in the supply chain, many types of software tools have been developed. These software programs are not the strategy, rather they are tools to implement a firm's strategy. The strategy is to focus the entire supply chain on satisfying the needs of the customer. Installing and using these tools is not the goal of the firm; the goal is to improve management of the supply chain. The information technology is an enabling technology that allows managers to do their job better because they have information that is more complete and more accurate than they would otherwise have.

There is a variety of software packages for each link in the supply chain. As computers and telecommunications equipment become cheaper, there will be even more advanced types of software available. To simplify the presentation of the types and role of the software used, software will be discussed here in 3 major sections. The internal linkages (software integrating our own firm's functions) will be discussed first, because this usually serves as the platform for integrating the firm with other software. Second, software that links our firm to our customers will be examined. Third, software that links our firm to our suppliers will be the final type that is reviewed.

Internal Data Integration

A common type of software used by firms to manage their internal portion of the supply chain is enterprise resources planning (ERP) system. An ERP system attempts to integrate all of the information processes in the organization and to use this integration to improve performance for the customer.

An enterprise resources planning system is a set of software modules that provides a company with the capability of automating the transactions involved with its business processes. The ERP system provides a common database and establishes uniform policies and practices across the entire enterprise. This allows real-time access to the data. An ERP system is an outgrowth of the traditional manufacturing software systems such as Material Requirements Planning Systems (MRPII). The traditional systems focused on planning and optimization of these plans. The ERP system expanded beyond these to serve additional functions in the firm.

ERP systems provide more data integrity, use of accessible databases, and consolidation of many different incompatible systems. The ERP system focuses on capturing all of the transactions in a firm. But, the ERP system does not suggest which decisions to make about the supply chain.

Enterprise resources planning (ERP) system—1) An accounting-oriented information system for identifying and planning the enterprise-wide resources needed to take, make, ship, and account for customer orders. An ERP system differs from the typical MRPII system in technical requirements such as graphical user interface, relational database, use of fourth-generation language and computer-assisted software engineering tools in development, client/server architecture, and open-system portability. 2) More generally, a method for the effective planning and control of all resources needed to take, make, ship, and account for customer orders in a manufacturing, distribution, or service company.

APICS Dictionary, 9th edition, 1998

Some software suppliers are developing interfaces for the ERP systems to interact with other ERPs across the web to allow Enterprise Supply Chain Management systems. These would address supply chain management and planning across an entire supply chain (i.e., the customer, suppliers, different divisions, etc.).

Advanced Planning Systems

Advanced planning systems (APS) are software programs that use data from the other systems which firms have to monitor their production and inventory. These other systems include ERP systems. The APS attempts to identify the best course of action for the firm. The APS creates production plans, schedules, and supply chain plans. The difficulty in implementing an APS is first to ensure coordination with the ERP or other system that performs that function. This ensures that the APS can obtain the needed data. The second issue is to change the organization structure and firm policies to use the output of the APS to obtain a competitive advantage. This may require that technology and the process be changed and will require training of the people involved.

The goal of an APS is to cut order cycle times and reduce inventories while increasing capacity use and throughput. It is essential that the APS have fast access to all available information to be able to accomplish this.

An APS can facilitate supply chain integration. It can track supplier capabilities and constraints and incorporate these into decisions. It can also track the capabilities of the transportation providers to ensure that distribution and delivery are not compromised.

The purpose of using an APS for supply chain management is to reduce over- and underproduction problems. One ability of most APS software packages is the ability to answer *what if* types of questions. This can be done using a simple simulation program. The software incorporates capacity constraints and financial goals into the planning process.

By using capacity more effectively, APS allows the firm to increase capacity utilization and to increase throughput if there are orders to use the capacity.

The APS does not replace ERP. Rather, it attempts to enhance the ERP system. An APS requires data from an ERP system to be able to calculate the delivery dates and material availability.

ERP and the Internet

Enterprise resources planning will not be killed by the Internet. To be useful on the Net though, ERP systems will require a front end that will allow interaction with customers and increased use internally. The fixes to this system will require an enormous effort.

It is here in the area of customer linkages that technologies seem to be advancing the most rapidly. Technologies that are simplifying the process of integrating the supply chain from the customer order to the supplier's product delivery include relational database system with real-time data, local area networks (LAN), wide area networks (WAN), and improved communications capabilities including the Internet and the World Wide Web.

Internet—A network of computer networks connected by means of telecommunication hardware that supports the global exchange of information.

APICS Dictionary, 9th edition, 1998

Local area network (LAN)—A high-speed data communication system for linking computer terminals, programs, storage, and graphic devices at multiple workstations distributed over a relatively small geographic area such as a building or campus.

APICS Dictionary, 9th edition, 1998

Wide area network (WAN)—A public or private data communication system for linking computers distributed over a large geographic area.

APICS Dictionary, 9th edition, 1998

The goal many firms have when installing this technology is to shorten the time it takes to receive an order, process it, prepare the product, and ship it to the customer. The purpose of shortening this order cycle is to increase customer service. To have the technology effectively reduce the order cycle requires that all appropriate personnel receive the necessary training so that they can turn the information obtained into knowledge and then act on this knowledge. Indeed, in one of its advertisements, SAP, the manufacturer of R/3, states that the rapid evolution of supply chain management is being driven by the integrated technology. It is this technology that enables companies to reengineer operations and link their entire business to align the flow of the goods with the market's demands. The emerging technology is increasingly capable of supporting interactive collaboration between all members in the supply chain.

The revolution in U.S. manufacturing philosophy that occurred after 1980 was to move from the *push* systems toward shop floor systems that *pull* materials into the process. The revolution occurring now is to extend this pull of materials from the customer throughout the entire supply chain.

For production to be able to respond to the customer's needs, it must know the customer's needs. Many firms are striving hard to produce and deliver their product very quickly to the customer so that they do not have to carry finished goods. Many attribute Dell Computer's phenomenal success to their ability to customize a customer's personal computer and deliver it quickly. Dell can do this because of the way it has organized its factory floor and the speed with which it can move information around the organization.

Whenever possible, firms want to use actual customer orders to schedule their factories. They prefer not to guess about potential demand. These firms do not schedule based on a forecast or a production plan because the shop's production lead time is shorter or equal to the length of time that a customer is willing to wait for a product. Their production process may be facilitated by flexible manufacturing environments which can process a wide range of volumes. And their administrative systems are supported by software which helps to quickly schedule the shop as the orders come in. In this environment, the customers may communicate on-line with the factory. This capability helps the firm replan quickly as the conditions change.

EDI has not progressed as rapidly as expected over the past decade. It was difficult to implement and it required intercompany standardization of how the transaction documents should be organized. Further there were often problems between points of connection. A typical EDI system is illustrated in Figure 17.1

Most EDI systems do not operate in real time. They batch process the data. The Internet makes it feasible to conduct exchanges between companies on a transaction-by-transaction basis.

While there are enormous potential savings from placing the entire supply chain online, it is difficult to achieve this. There are technological, logistical, and conceptual barriers to accomplishing this. In the auto industry, the Big 3 began electronically linking suppliers before 1985 with the goal of improving the flow of materials. That was to be just the first step, while later steps would incorporate engineering design systems and financial transaction systems. As part of this, the Automotive Industry Action Group (AIAG founded in 1981 to develop specific product-oriented standards) established industry-wide electronic communication standards in the mid-80s. Great progress has been made. All or most of the tier 1 suppliers electronically communicate purchase orders and releases for production, but the more sophisticated data is not shared electronically and few of the tier 1 suppliers are linked electronically with their own suppliers.

One barrier has been the incompatibility of the different software systems used by the OEM and its suppliers. Further, many engineers prefer working with 2-dimensional figures on paper to looking at 3-dimensional files on the computer. Another barrier is that even some of the suppliers who have EDI capability do not incorporate it into their overall manufacturing and

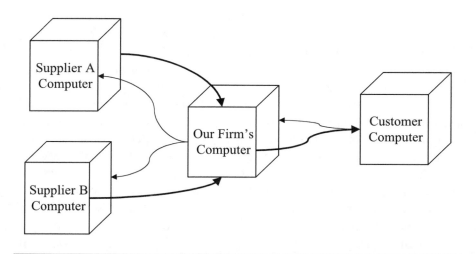

Figure 17.1   Electronic Data Interchange (EDI) in Business-to-Business Communication

inventory-management information systems. The information comes in on one system, and then it is keyed into second system.

Electronic Commerce on the Internet

The Internet can help companies lower costs throughout their supply chain. It is also possible to use the Internet to improve customer service. Industry is just starting to use the Internet, but as industrial leaders begin to use it they will force others in their industry to do the same thing. Both large firms and small firms will embrace the Internet over the next few years. The computer industry has already done so. As the use of the Internet allows decreased costs and increased customer contact, these firms will become more competitive and put increasing pressure on their competition to take the same steps.

What firms who are leaders in the Internet applications are striving to achieve is the transfer of many of their core processes to the Web. This means that firms will need Websites that can support online transactions and can share data with their customers' databases and their suppliers' database. Past investments in information technology have primarily been focused on improving the internal operations of the firm. The Internet allows increased communication and connection with the outside world. Electronic business allows the benefits of speed and automation of internal processes to be shared with customers and suppliers. The ability to collaborate with others is an additional competitive advantage.

The advantage of the Internet over EDI is that it is much more flexible. The Internet can adapt quickly to changes in market conditions. The Internet also has open standards.

Companies can use the Internet to facilitate communication internally on what is referred to as an intranet. The information that would sit in logistics, or manufacturing or finance, etc., can be shared quickly between functions. This increases the amount of internal collaboration that is possible.

Collaboration can also be increased with suppliers and customers by creating an extranet. In an extranet, corporate information is shared with selected outsiders. This allows collaboration and cooperation with the firms outside. It is hoped that suppliers will match their production patterns to the demand patterns they can observe as the demand information is shared.

Barriers to Using the Internet in Supply Chain Integration

The major barrier to increased supply chain integration is the mind set of the managers of the different firms involved. With or without an Internet, most managers view their firm as being an independent system. They do not have a systematic view of the supply chain. Given this attitude, it is difficult for a firm to develop the openness necessary to share information freely over an extranet.

Cisco Systems outsources all of its manufacturing. It has its suppliers post their quotes and forecasts on Cisco's Website each quarter. Then Cisco selects its suppliers and contract manufacturers (*Economist,* June 26, 1999).

A second barrier to integrating the Internet into a company is the need for services in addition to products. This is an opportunity as well as a challenge. Firms will have the opportunity to sell services as well as a product to customers on the Net. The challenge is that many of the services will not be provided by the firm selling the product. Instead, the service may be outsourced to a partner. For example, one service provided by the Home Depot chain of building supply stores is that contractors can be given account numbers. The contractor can use that account number to log onto Home Depot's Website. The Website will provide services such as material estimates and work schedules for jobs. It will also allow the contractor to schedule deliveries.

Dell Computer's Factory

Dell sends orders to its factory in Ireland (responsible for manufacturing PCs for Europe) from its Website and call centers. This information is relayed to its suppliers, so that they know the components needed and when they are needed. As the components enter, they are assembled and the final assembled computer is shipped a few hours later.

Dell sells over $15 million of computers from its Website each day. Dell's suppliers have real-time access to information about orders that Dell receives over the Internet. Dell also allows customers to track the progress of their order from the factory to their doorstep on the Internet.

MIS Implementation

There are many success stories and failures in the business press about information technology implementations. The first item that is necessary for a successful implementation is that the firm have a clear vision and mission of what it is trying to accomplish. This allows the firm to have a clearly stated strategy, which then allows all the managers to understand how a new information technology system can contribute to achieving the strategic goals of the firm.

It is only by being able to write clear, detail definitions of the system requirements that a firm can hope to obtain a good information technology system. Managers must be good users of the MIS system to implement a good system. One of the requirements is that they forecast the needs of the company for the future as well as the current needs.

Success requires coordination among the different levels of management. Unless the firm is small, the day-to-day interface with the system will be by lower level managers. They must be able to clearly state their requirements of the system. It is only in this way that a solid list of deliverables can be developed. As the software is being developed, a firm must also consider the training needs of the staff who will use the software. Is there documentation available for them, that they will be able to understand? Will the training be online, or will there be a need to hire trainers? Is it possible to develop train-the-trainer programs? Another very important item is to ensure that the data entered into the system is accurate. There is always the possibility of clerical error, but there is also the issue of whether the existing procedures ensure accuracy now. If there are no checks to ensure that the invoice or the shipping order received is accurate now, what will need to be put into place to achieve accuracy after the new system is implemented?

Once implementation is complete (i.e., the software is installed, in use, and actually works), there is a continued need for the implementation process to continue. All the staff who use the system need to be retrained and any problems resolved. The training sessions themselves only introduce people. The actual training is on-the-job training, as everyone tries to accomplish his/her job using the system.

18 Summary

Although *supply chain management* continues to be a subject much in the conversations of business leaders and often the topic of various industry magazine articles, it remains a largely misunderstood and often misapplied business philosophy. In its most basic sense, the supply chain is nothing more than a holistic view of the business enterprise from the origin of raw materials through to the use of the completed product or service by the ultimate consumer.

However, as fundamental as this concept is and as intuitive as optimization of the supply chain is, many companies still attempt to maximize only that which they specifically control. They overlook the value of the global view. They concentrate only on what they can see in their local operation. The idea of real supply chain optimization is looked upon as "something to be done," some objective for the future. "Once we get our own act in order, we can move on to the supply chain issues" is a common statement from business leaders in every industry. In some cases, when leaders do apply some effort toward supply chain optimization, it is only half-hearted and does not receive the attention or financial budget that it deserves. Those companies, even more so as the Internet makes global business operations a reality, are not only missing some opportunities, they are sealing their fate as noncompetitive businesses in the modern age.

Effective supply chain management is no longer an option; it is a requirement for survival. Companies within the supply chain must reach new levels of communication and cooperation. Rather than treating each other as adversaries and attempting to gain competitive advantages at the expense of each other, companies with effective supply chain strategies are able to break

223

the traditional paradigms. To survive in today's competitive business climate, organizations must take a broad view of their product and/or service flow. They must consider that the opportunities for system optimization offer much greater potential to bottom-line business results than can be obtained from minor efficiency improvements that stem from the more traditional approach to local business management. Supply chain management allows a much more proactive approach to the typical issues facing businesses in the 2000s. It allows managers the chance to see the impact of local decisions within any element of the entire supply chain on the global results of that chain. Conversely, not practicing these supply chain philosophies and not assuming this global view will render those companies noncompetitive. Those who practice effective supply chain management techniques will surpass and ultimately defeat those who do not.

The key enabler to accomplish this is technology. Fortunately, the technology tools needed to help manage an entire supply chain are abundantly available. Transportation and communication tools are being upgraded daily such that seemingly established paradigms are being broken at alarming rates. Advanced planning and scheduling (APS) systems are also in rapid development offering unprecedented power in the hands of supply chain managers to use complex algorithms to generate detailed plans for all parts of the chain. Simulation modeling software allows strategic *what if* analyses to be completed within minutes, greatly reducing the risk of some bold decisions that otherwise might have been rejected.

Business can do more now with electronic technology than could even be imagined only a few years ago. And, the rate of development promises to increase. Certainly, those tools will make managing a complex supply chain ever more possible. However, the danger exists that business leaders will assume that the tools in and of themselves will provide all the results that they seek. This is simply not true. The technology, no matter how advanced it will become, will forever be only a tool and will not accomplish the synchronization of the supply chain without significant behavioral change throughout every element of the system. How elements of the chain work with each other through real-time communication and coordination will make the difference. Technology is only a tool. How we use that technology to help coordinate the efforts of the entire supply chain will determine its value.

The new paradigms concerning supply chain management involve total integration such that each element of an entire supply chain looks through its supplier's supplier and its customer's customer. This overlapping of vision will create a much more integrated view of the supply chain for every element.

Technology will allow all upstream and downstream trading partners to share everything from strategic plans to operational data through electronic linkages.

The primary driving force toward supply chain management results from the ever higher demands of the consumer toward quality, service, and competitive pricing. Clearly, the ultimate goal of any company is to make money. That is accomplished by providing a quality product or service to the customer within the time frame in which it is needed and doing all of that at a reasonable price which is acceptable to the customer. By coordinating the creation elements of that product or service within the entire supply chain as described in this book, any organization will dramatically improve its chances to make money now and in the future.

The essence of supply chain management is *communication,* allowing the ultimate consumer to become a partner in the process. As a result, not only is the product or service more likely to be delivered to the customer as it is currently desired, but also the product or service can actually change in design as the needs of the customer changes. This close and real-time communication linkage, created through the practices described in this book, helps all the elements of the supply chain to meet the ever changing demands of the ultimate consumer. And, those companies that learn to constantly meet those demands are the companies that will continue to meet their own goals of making money. This is the ultimate effect of an effective supply chain management strategy and practice.

References and
Further Readings

Bhaskaran, S., "Simulation Analysis of a Manufacturing Supply Chain," *Decision Sciences,* 29 (3), 633–657, 1998.

Bidault, F. and C. Butler, "Early Supplier Involvement: Leveraging Know-how for Better Product Development," *Target: Innovation at Work,* 12 (1), 20–26, 1996.

Burdett, R. L. and G. W. Cleaves, "Integrated Sales Manufacturing and Logistics," *APICS—The Performance Advantage,* 6(8), 62–66, 1996.

Chaudry, O., "Enterprise Supply Chain Management," *APICS—The Performance Advantage,* 8 (9), 46–48, 1998.

Cohen, S. G. and D. E. Bailey, "What Makes Teams Work: Group Effectiveness Research from the Shop Floor to the Executive Suite," *Journal of Management,* 23 (3), 239–290, 1997.

Copacino, W. C., *Supply Chain Management: The Basics and Beyond,* Boca Raton, FL: St. Lucie Press, 1997.

Covington, J. W., *Tough Fabric,* Severna Park, MD: Cheasapeake Consulting, 1996.

Cox, J. F., III and J. H. Blackstone, Jr., *APICS: Dictionary,* 8th Edition, Falls Church, VA: American Production and Inventory Control Society, Inc., 1995.

Cox, J. F., III and J. H. Blackstone, Jr., *APICS: Dictionary,* 9th Edition, Falls Church, VA: American Production and Inventory Control Society, Inc., 1998.

Dobler, D. W., L. Lee, Jr., and D. N. Burt, *Purchasing and Materials Management,* 4th Edition, New York: McGraw-Hill, 1984.

Doyle, M. F., "Fundamentals of Strategic Supply Management," *Purchasing World,* 40–41, December 1990.

Evans, J. R. and W. M. Lindsay, *The Management and Control of Quality,* Cincinnati, OH: South-Western College Publishing, 1999.

Finch, B. J. and R. L. Luebbe, *Operations Management: Competing in a Changing Environment,* New York: Dryden Press, 1995.

Flynn, B. B., R. Schroeder, and S. Sakakibara, "Determinants of Quality Performance in High- and Low-Quality Plants," *Quality Management Journal,* 2(2), 8–25, 1995.

Forker, L. B., D. Mendez, and J. C. Hershauer. "Total Quality Management in the Supply Chain: What Is Its Impact on Performance?" *International Journal of Production Research,* 35 (6), 1681–1701, 1997.

Gable, R., "The History of Consumer Goods: How Supply-Chain Management Is Driving the Next Consumer Goods Revolution," *Manufacturing Systems,* 15 (10), 70–84, 1997.

Goldratt, E. M. and J. Cox, *The Goal,* 2nd Revised Edition, Croton-on-Hudson, NY: North River Press, 1992.

Goodman, P. S., *Designing Effective Work Groups,* San Francisco: Jossey-Bass, 1986.

Grieco, P. L., Jr., *Supplier Certification II: Handbook for Achieving Excellence through Continuous Improvement,* 5th edition, West Palm Beach, FL: PT Publications, Inc., 1996.

Hahn, C. K., C. A. Watts, and K. Y. Kim, "The Supplier Development Program: A Conceptual Model," *Journal of Purchasing and Materials Management,* 2–7, Spring 1990.

Hall, R. W., *Zero Inventories,* Homewood, IL: Dow Jones–Irwin, 1983.

Handfield, R. B. and Nichols, E. L., Jr., *Introduction to Supply Chain Management,* Upper Saddle River, NJ: Prentice-Hall, 1999.

Harwick, Tom, "Optimal Decision-Making for the Supply Chain." *APICS—The Performance Advantage,* 7, 42–44, 1997.

Hayes, R. H. and S. C. Wheelwright, "Link Manufacturing Process and Life Cycles," *Harvard Business Review,* 19779, 133–140, January–February 1979.

Hayes, R. H. and S. C. Wheelwright, *Restoring Our Competitive Edge: Competing through Manufacturing,* New York: John Wiley & Sons, 1984.

Hill, S., Jr., "We Need to Talk: Collaboration Is the Next Step to Supply-Chain Improvements," *Manufacturing Systems,* 16, 40–48, 1998.

Hill, T., *Manufacturing Strategy,* Boston, MA: Irwin, 1989.

Imai, Masaaki, *Kaizen,* New York: McGraw-Hill, 1986.

Ingalls, J., "How Design Teams Use DFM/A to Lower Costs and Speed Products to Market," *Target: Innovation at Work,* 12 (1), 13–19, 1996.

Ishikawa, K., *Guide to Quality Control,* White Plains, NY: Asian Productivity Organization: Quality Resources, 1982.

Jacobs, F., D. Robert, and C. Whybark, *Why ERP? A Primer on SAP Implementation,* Boston, MA: Irwin, 2000.

Juran, J. M. and F. M. Gryna, Editors, *Juran's Quality Control Handbook,* 4th Edition, New York: McGraw-Hill, 1988.

Killen, K. H. and J. W. Kamauff, *Managing Purchasing: Making the Supply Team Work,* New York: McGraw-Hill, 1995.

Kilpatrick, J., "Advanced Planning Systems Spark the Supply Chain," *APICS—The Performance Advantage,* 9 (8), 24–28, 1999.

Krause, D. R. and L. M. Ellram, "Critical Elements of Supplier Development: The Buying-Firm Perspective," *European Journal of Purchasing and Supply Management,* 3 (1), 21–31, 1997.

Krause, D. R., R. B. Handfield, and T. V. Scannell, "An Empirical Investigation of Supplier Development: Reactive and Strategic Processes," *Journal of Operations Management,* 17, 39–58, 1998.

Lamprecht, J. L., *ISO 9000: Preparing for Registration,* Milwaukee, WI: ASQC Quality Press/Marcel Dekker, Inc., 1992.

Lamprecht, J. L., *ISO 9000: Implementation for Small Business,* Milwaukee, WI: ASQC Quality Press, 1996.

Larson, C. E. and F. M. J. LaFasto, *Team Work: What Must Go Right/What Can Go Wrong,* Newbury Park, CA: Sage Publications, 1989.

Lawler, E. E., *The Ultimate Advantage: Creating the High-Involvement Organization,* San Francisco: Jossey-Bass, 1992.

Layden, J., "The Reality of APS Systems," *APICS—The Performance Advantage,* 8 (9), 50–52, 1998.

Leenders, M. R., H. E. Fearon, and W. B. England, *Purchasing and Materials Management*, 9th Edition, Homewood, IL: Irwin, 1989.

Maack, R., "Russell Stover Looks to Optimizing Supply Chain to Sweeten Bottom Line," *APICS—The Performance Advantage*, 2–3, October Supplement, 1998.

Melnyk, S. A. and D. Denzler, *Production Operations Management*, Chicago, IL: Irwin, 1996.

Mills, C. A., *The Quality Audit: A Management Evaluation Tool*, Milwaukee, WI: McGraw-Hill, 1989.

Mohrman, S. A., S. G. Cohen, and A. M. Mohrman, Jr., *Designing Team-Based Organizations: New Forms for Knowledge Work*, San Francisco, CA: Jossey-Bass, 1995.

Monden, Y., *Toyota Production System: An Integrated Approach to Just-in-Time*, 2nd Edition, Norcross, GA: Industrial Engineering and Management Press, Institute of Industrial Engineers, 1993.

Parker, Kevin, "Manufacts: Detroit a Hot-Bed for E-Commerce," *Manufacturing Systems*, 18, June 2000.

Poirier, C. C. and Reiter, S. E., *Supply Chain Optimization: Building the Strongest Total Business Network*, San Francisco: Berrett-Koehler Publishers, 1996.

Schneiderjans, M. J., *Topics in Just-in-Time Management*, Boston, MA: Allyn and Bacon, 1993.

Schults, G., "Rapid Response Is a Strategy, Not an Application," *Managing Automation*, 12, (4), 32–35, 1997.

Senge, P. M., *The Fifth Discipline*, New York: Doubleday, pp 27–54, 1990.

Shiba, S., A. Graham, and D. Walden, *A New American TQM: Four Practical Revolutions in Management*, Portland, OR: Productivity Press, 1993.

Simmons, B. L. and M. A. White, "The Relationship Between ISO 9000 and Business Performance: Does Registration Really Matter?" *Journal of Managerial Issues*, 11 (3), 330–343, 1999.

Steinbacher, H. R. and N. L. Steinbacher, *TPM for America: What It Is and Why You Need It*, Portland, OR: Productivity Press, 1993.

Sugiyama, T., *The 5S Approach to Improvement*, New York: PHP Institute, 1995.

Tajiri, M. and F. Gotoh, *TPM Implementation: A Japanese Approach*, New York: McGraw-Hill, 1992.

"The Net Imperative," *The Economist: A Survey of Business and the Internet*, 351 (8125), 5–40, 1999.

Thondavadi, N. N. and A. Raza., "Facilitating ISO 9000 Certification Teams," *APICS—The Performance Advantage*, 8 (9), 40–44, 1998.

Vasilash, G. S., "The Road Less Traveled," *Automotive Manufacturing and Production*, 110 (6), 8, 1998.

Wageman, R., "Interdependence and Group Effectiveness," *Administrative Science Quarterly*, 40, 145–180, March 1997.

Winter, D., "The Paper Chase: Trees Still Dying in Auto Industry's Electronic Age," *Ward's Auto World*, 34 41–42, June 1998.

INDEX